T0186996

MATHEMATICAL
INTELLIGENCE

MATHEMATICAL INTELLIGENCE

What we have that machines don't

JUNAID MUBEEN

P

PROFILE BOOKS

First published in Great Britain in 2022 by
Profile Books Ltd
29 Cloth Fair
London
EC1A 7JQ

www.profilebooks.com

A CIP catalogue record for this book is available from the British Library.

ISBN 978 1 78816 683 6
eISBN 978 1 78283 795 4

Typeset in Sabon by MacGuru Ltd
Printed and bound in Great Britain by Clays Ltd, Elcograf S.p.A.

For Leena,
my pride *and* my joy.

Contents

ACKNOWLEDGEMENTS

'Acknowledgements' are inadequate. Every person mentioned here has my heartfelt gratitude for helping me turn a vague concept into an actual thing. Any shortcomings are my own.

My agent, Doug Young, elevated my ambitions for what this book could become. He has been a strong advocate and guiding hand throughout.

I'll never forget my first conversation with Helen Conford because it was the first time I felt validated as an aspiring author. I'll always be grateful for the chance she took in bringing me to Profile. Ed Lake, Paul Forty and the whole editorial team have worked their magic to polish and fine-tune the rough manuscript that was presented to them.

Several friends and colleagues vetted early drafts: my thanks to Keith Devlin, Shameq Sayeed, Steve Buckley, Roxana and Rares Pamfil (the golden couple!), Noel-Ann Bradshaw, Andrew Mellor, Lucy Rycroft-Smith, David Seifert and Ed Border. A special mention to Mohamady El-Gaby for sense-checking my neuroscience claims and helping me find my footing as I ventured far beyond my own areas of expertise. Thanks also to Taimur Abdaal for lending me a sketch or two.

Much of this book was written on Friday afternoons in coffee shops and I'm thankful to my friend and former boss,

Richard Marett, for granting me '10% time' to indulge in this project during my time at Whizz.

I have seen mathematical intelligence in action every Sunday in the Oxford Maths Club. To every parent who has entrusted me with their child's maths development, and to every student who has taken to our courses with gusto, I am immeasurably grateful.

My wife, Kawther, is the unsung hero of this project. She has championed my work even when it was little more than scrawlings on a napkin. She also happens to be the literary talent in the family, and a ruthless editor to boot. Since the inception of this book we've grown ourselves a delightful family. Leena and Elias are my two greatest blessings in life; the book is dedicated to the former (next one's for you, *lumps*).

THE CASE FOR MATHEMATICAL INTELLIGENCE

MIT, 1950s. The first wave of Artificial Intelligence is on the horizon. Marvin Minsky, one of the field's leading figures, proclaims: 'We're going to make machines intelligent. We're going to make them conscious.' Douglas Engelbart, a peer of Minsky's, retorts: 'You're going to do all that for the machines? What are you going to do for the people?'[1]

Artificial intelligence (AI) researchers are nothing if not bullish about the prospects of their creations. The field kicked off in earnest in 1956 at a summer workshop held at Dartmouth College, New Hampshire, where the founding fathers of AI set out their vision in no uncertain terms. Intelligent machines, they believed, were to propel humanity into the next golden age of innovation by 'simulating every aspect of learning or any other feature of intelligence'.[2] The timeframe was bolder still: one summer was all they would need to break the back of AI.

Things turned out to be rather more complicated, as a summer of hype gave way to a succession of AI winters, with progress in the field largely stagnant for several decades. But if you've caught the headlines recently, you'll know that AI is currently the subject of renewed hype. Between flagship triumphs in popular games, the growing presence of home assistants, and the coming of self-driving cars, the machines have resumed their rise.

We humans have distinguished ourselves from other species by inventing tools to help us solve our most challenging problems. And yet we may be complicit in our own demise because some of these tools have become so powerful that they appear to pose genuine threats to our ways of thinking and being. Studies of the growing threat of automation to human labour abound,[3] while the so-called 'superintelligent' machines of tomorrow may force us to re-examine what it even means to be human in the first place.

As we enter this new cycle of ratcheted expectations, hopes and anxieties around the latest wave of technological innovation, Engelbart's question should resonate loud and clear. We reserve such reverence for technology that we risk overlooking our own human capabilities. Machines lack some of the basic qualities of human thinking – qualities we have sidelined through our mechanistic ways of schooling and working, and qualities that we need to urgently reawaken to thrive alongside our silicon counterparts.

As it happens, humans have – through millions of years of evolution and thousands of years of continual refinement – developed a powerful system for making sense of the world, for imagining new ones, and for devising and solving complex problems. This system has helped us create the economies that underpin our society. It has shaped our notions of democracy. It has spawned technologies that now stare us down, but the same system can equip us with the skills to tame these digital beasts.

The system has a name: *mathematics*.

What is mathematics, really?

Mathematics has been described as an art, a language and a science. For some, it is a means of unlocking nature's secrets. As Galileo testified so eloquently: '[The universe] cannot be read until we have learnt the language and become familiar with the characters in which it is written. It is written in mathematical language.' This is mathematics as the language of the universe, the engine of scientific progress.

The scope of mathematics transcends our physical universe. Entire swathes of the subject are explored for their own sake, driven by the deep satisfaction that comes from dreaming up new concepts, piecing together ideas, and grappling with thorny problems. Many mathematicians seek out aesthetic qualities in their craft. The twentieth-century mathematician and philosopher Bertrand Russell spoke of the subject's 'supreme beauty – a beauty cold and austere ... capable of a stern perfection such as only the greatest art can show'.[4] Many see themselves as artists as well as scientists – 'makers of patterns',[5] to borrow a description from G. H. Hardy, a contemporary of Russell's. It is not uncommon for mathematicians to deride the need to apply their thinking to the 'real' world, as if utility were some kind of distraction. It has even been proposed that some aspects of mathematical inquiry have a hedonistic basis.[6]

From these varied motives, mathematics is often partitioned into two supposed types: there is *applied* mathematics, which, as the name suggests, is concerned with problems of the real world. Then there is the presumptively labelled strand of *pure* mathematics, which centres on more abstract concepts and rigorous arguments often removed from practical consideration. This separation is felt keenly at university, where maths students are expected to declare their allegiances before specialising in one area. I was of the *pure* persuasion. Yet, since leaving formal mathematics a decade ago, much of my work has been rooted in datasets and algorithms – about as *applied* as it comes.

Having bridged the pure/applied divide, I have come to realise that it is an arbitrary and limiting way of characterising the subject. There is a commonality that binds mathematicians of all types. Without exception, we derive immense joy from tackling maths problems, a satisfaction akin to solving our favourite puzzles. Mathematics is even alleged to elicit the same physiological reactions as sexual activity (yes, really).[7] Alongside that pleasure comes power; whatever branch of mathematics a mathematician happens to be probing, they are using the mind's highest faculties and building a store of portable mental models that serve them in all parts of life.

It may feel risky to invest time and effort in studying mathematics based on nebulous notions of pleasure and power. But mathematics cannot help but bring practical uses too. It is not at all unusual for a field of mathematics that starts out as pure intellectual inquiry to later find itself in practical settings. Prime numbers (whole numbers greater than 1 that cannot be divided into smaller whole number parts) were first studied for their unusual arithmetic properties, yet internet security now relies on them – your credit card details are kept secure by the sheer difficulty of finding the prime factors of really large numbers. The Greeks were enthralled by the geometric properties of ellipses; only centuries later would Kepler discover that planets move around the sun in an elliptical orbit. The topology of knots, a delight to study in its own right, has applications in protein folding. And calculus (the study of continuous change), arguably the most applied of all mathematical topics, which was the basis for Newton's study of planetary motion, and whose tools are indispensable to engineers, physicists, financial analysts, even historians,[8] was developed within the rigorous frameworks of pure mathematics. I could go on.

The theoretical physicist Eugene Wigner encapsulated this entwining of intellectual curiosity and utility by remarking on what he called the 'unreasonable effectiveness' of mathematics, declaring that 'the enormous usefulness of mathematics in the natural sciences is something bordering on the mysterious and there is no rational explanation for it'.[9]

The 'usefulness' of mathematics is not limited to specific real-world applications. It arises chiefly from its invitation to explore a vast range of concepts, even arcane ones. Mathematics transports us into multiple worlds, each governed by its own rules. It encourages us to break free of convention and leap from one conceptual system to another. These faraway worlds can also train us to think in ways that enrich our understanding of our own, physical one. Even as the content of my own doctorate in pure mathematics drifts from memory (to the point where I can scarcely grasp its essential ideas any longer),[10] the process by which it was created remains its most enduring contribution to my everyday thinking and problem solving.

Mathematical intelligence is not about calculus or topology any more than musical intelligence is limited to a particular genre or instrument. It is a system for making us better thinkers and problem solvers, using the proven tools of mathematicians. And in the age of smart machines, it is needed more than ever.

Mathematics and calculation: a false coupling

The mathematics I've just described is quite apart from what we encounter at school. 'School mathematics' places great emphasis on calculation. A calculation is a routine operation performed on certain objects, often numbers, to produce a particular result. It can be as simple as counting and as complicated as Google's search-ranking algorithms (an algorithm here just means a list of step-by-step instructions).* School mathematics is premised on the idea that rehearsing a litany of routine calculational techniques is a strict prerequisite for mathematical intelligence, and a gateway to employment. Topics such as calculus, algebra and geometry, each of which contain a multitude of rich concepts, are stripped down to bare calculational form.

The marriage between mathematics and calculation is the result of several forces. The first is an industrial paradigm of formal education whose roots can be traced back to the mid-nineteenth century, when the aims of mass schooling coalesced with notions of mechanisation and scale, and increases in urban populations fuelled demand for everyday numeracy skills such as counting money and telling the time. As universal education systems sprang up the world over, subject matter reflected the needs of a mathematically literate workforce. In England, for instance, arithmetic dominated the curriculum, and additional topics – such as algebra, mechanics and fractions – were introduced with the goals of employment in mind.[11]

Society has made giant leaps of progress since then, yet school mathematics has remained largely static; national and international curriculum standards remain heavily couched in speed and proficiency with calculation. The stubborn persistence of calculation in education also owes a debt to widely held beliefs around the nature of mathematics. *Platonism*

---

*Computation and calculation have slightly different meanings. The former tends to refer to algorithmic processes, the latter to arithmetic ones. I will use them interchangeably because they both espouse the same kinds of routine thinking processes.

– first espoused by the Greek philosopher Plato – holds that mathematical objects are abstract entities independent of language, thought or practices. Just as electrons and planets exist independently of us, so do mathematical concepts such as number. In this view, there is a single form of mathematics, timeless and immutable. Alongside Platonism, there is the *formalist* view, which gained traction in the twentieth century and considers mathematics a self-contained system of logical truths, each derivable from first principles. The Platonist and formalist philosophies, especially popular among 'pure' mathematicians, conspire to reduce mathematics to a single pathway of predetermined, hard-coded truths. Abstraction is the gold standard in this framing of mathematics, its raison d'être, best accessed by mastering symbol manipulation. The execution of mathematical procedures – fast, precise calculation – is seen as the single pathway to deep mathematical thought.

The Platonist–formalist view overlooks the crucial fact that mathematics takes on rich and diverse forms,[12] all of which are birthed in the context of local environment and experience.[13] Take something as seemingly universal as our number system. It arises out of a series of choices, from the symbols we use to denote quantities to how we group together objects to manage large amounts, to how we perform arithmetic on numbers. In schools across the world, students are taught Hindu–Arabic numerals (0, 1, 2, and so on), the decimal system (grouping numbers into tens), and specific algorithms for performing addition, subtraction, multiplication and division. Students are led to believe that these choices are inevitable – the only conceivable way to think about numbers – when in fact they are situated within a historical and sociocultural backdrop. As we'll see in later chapters, communities around the world to this day adopt highly varied representations for numbers.

Mathematics in the real world is more situational and contextual than Platonism and formalism would suggest.

My work has taken me to classrooms the world over and I can confirm that, despite its short-sightedness, the Platonist–formalist ideal is alive and well everywhere. A common thread binds the mathematics taught to marginalised communities in Kenya, children of Microsoft executives in Washington State, students of Eton College, and low-income families in rural Mexico. In all these cases, school mathematics is characterised by a heavy diet of calculation,[14] and mathematical talent is conceived as the ability to execute these techniques flawlessly and at speed.

School mathematics comes wrapped in the promise that this very particular skill set will, on some unspecified date in the future, serve students' everyday needs. That promise may have held up in the nineteenth century, when, for example, the formulae of trigonometry would guide your career as a carpenter or surveyor or navigator, and you would be expected to make the requisite calculations by hand. Yet the twenty-first-century student will discover, if they haven't already, that calculation is no longer the unique marker of human mathematical talent. It is almost tautological to say it, but for computation we have computers.

School mathematics is clearly in need of a rethink, which should come as welcome relief to most. Far from evoking the sentiments of wonder or beauty experienced by mathematicians, it is more commonly associated with feelings of dread. In the UK alone, a fifth of the population is afflicted with maths anxiety.[15] For these people, the anticipation and experience of doing mathematics activate the same regions of the brain that give rise to pain.[16] Attitudes towards mathematics have been shown to deteriorate with age,[17] and many people, scarred by their encounters in school, flee into the safe sanctuary of adulthood, resolving never again to confront anything that resembles mathematics. Is the Platonist–formalist method of education simply the price we have to pay to feel the power of mathematics – to appreciate its unreasonable effectiveness? Even if a casualty rate of one in five is deemed palatable, the apparent victors of this brand of mathematics find themselves trapped in a false sense of security. As an admissions tutor at Oxford University, and more recently as an employer, I have interviewed hundreds of candidates who naïvely presume that a clean sweep of top grades in mathematics at school has prepared them to think creatively and tackle complex problems.

The German poet Hans Magnus Enzensberger has described mathematics as 'a blind spot in our culture – alien territory, in which only the elite, the initiated few have managed to entrench themselves'.[18] There is a yawning chasm between the mathematics enjoyed by professional mathematicians and the monotony of most school curricula.

Professional mathematicians, for their part, tend to keep calculation at arm's length. They recognise that techniques such as long division, the quadratic formula and trigonometric identities occupy a small space within the mathematical landscape, a tiny sliver of all the concepts available in the subject. Entire branches of mathematics are removed from calculation, and even where calculations surface, the creative elements of mathematical intelligence reside in dreaming up such methods in the first place, understanding their inner workings and applying them in novel settings. The specific act of calculation is secondary and offers little joy or illumination.

New calculating tools, new mathematics

The history of mathematics shares a timeline with an ongoing effort to liberate humans from the tedium of calculation. Performing calculations does not come naturally to us. Time and again, we have created tools and technologies that outsource the most mechanical aspects of mathematics.

Great leaps have been made with leading-edge calculating tools.[19] Where our earliest ancestors marshalled pebbles and grains to keep track of basic quantities, the city planners of Babylonia, Sumeria and Egypt used formal calculation schemes which were brought to bear on problems of engineering, land administration, astronomy, timekeeping, planning and logistics. Calculation, along with reading and writing, became a cornerstone of more developed civilisations. Some of the earliest surviving government records are replete with calculations central to administration.

Physical counting instruments were always close at hand. The abacus that helps us count large quantities has its roots in the pebble-counting schemes of the ancient Romans, and as calculations grew in complexity, so too did the power of our tools. Older readers may recall using a slide rule in school to assist in weighty calculations such as the multiplication of large numbers. The slide rule was based on John Napier's logarithm tables. Napier was born into a Scottish family of estate owners in 1550. Copernicus had just developed the heliocentric model of the universe, placing the sun at its centre for the first time. Columbus had sailed the Atlantic, and Renaissance artists were advancing their own frontiers. Yet the world remained heavily dependent on tired calculational conventions. The work of masons, merchants, navigators and astronomers all required methods of long division and multiplication that were tediously handcrafted, prone to human error, and prohibitively expensive to carry out (pen and paper did not come cheap).

On his travels across Europe as a young student, Napier observed the burden of calculation first-hand. He would encounter decorative books composed solely of mathematical tables and currency versions, created and used daily by merchants. The tables still demanded a hefty degree of calculation on the part of the user. There had to be a more effective way, Napier thought, of removing what he called 'those hindrances' to trade. Napier was alluding to what cognitive psychologists now term our 'working memory', which handles short-term information and is limited to between four and seven objects at a time.[20] This makes multistep calculations such as long multiplication or division difficult to perform, as we strain to keep track of each moving part.

In his famous work *Mirifici Logarithmorum Canonis Descriptio* ('Description of the marvellous canon of logarithms'), Napier introduced a powerful mathematical object called the logarithm function. To grasp the intuition behind logarithms, first consider a familiar multiplication involving powers of 10:

$$\underbrace{100}_{\substack{2 \\ \text{zeros}}} \times \underbrace{1000}_{\substack{3 \\ \text{zeros}}} = \underbrace{100000}_{\substack{5 \\ \text{zeros}}}$$

This calculation is straightforward because we just 'add the zeros' in each term to get our product. It would be handy if every multiplication could be managed in such a simple way. Napier's logarithm makes this possible. In the numbers above, the string of zeros corresponds to how many times 10 multiplies by itself – twice for 100, thrice for 1000, and so on. With this in mind, the logarithm of a number is defined by how many times you have to multiply 10 by itself to get that number. So the logarithm of 100, denoted log(100), is 2, and the logarithm of 1000, denoted log(1000), is 3.

The clever, mathematical part is that the logarithm can be defined for every positive number, not just powers of 10. The logarithm of 95 is 1.978, the logarithm of 2367 is 3.374, and the logarithm of 3 is 0.477, which is to say that 'if you multiply 10 by itself 0.477 times you will get 3'. That may sound strange at first, but the conceptual power of mathematical functions allows us to bring such notions into existence.

A useful property of logarithms is that they obey the following rule:

$$\log(a \times b) = \log(a) + \log(b)$$

Suppose we want to multiply two large numbers. Napier explained how, using the above formula, we can transform the problem into one involving addition, which is simpler and less error-prone. All we need is a table that lists the 'logarithmic value' of each number. The process then goes as follows:

1. Look up the logarithm of each value to be multiplied.
2. Add these two logarithmic values to get a total.
3. Look up the value of the number whose logarithm corresponds to this total. The number you have found is the product of your two original numbers.*

---

*This method is a slight simplification of how Napier's tables were constructed, but close enough to give reasonable approximations, which is often all we need. It uses the *base 10* logarithm, which can be substituted for any other value – the natural logarithm that is now popular calls on base *e*, where *e* denotes Euler's number.

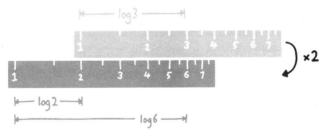

A slide rule in action: if we slide the top ruler 2 units along the ruler below (a length of log 2) then every number on the ruler below corresponds to multiplying the number above it by 2. For example, the number 3 (which is a length of log 3 along the top ruler) lines up with the number 6 (which is a length of log 6 along the bottom ruler), telling us that 3 × 2 = 6.

Napier's *Canon* comprised huge lists of numbers and their corresponding logarithmic values. It took some twenty years to compile. When dedicating the work to the future King Charles I, Napier wrote of how 'this new course ... doth clean take away the difficulty that heretofore hath been in mathematical calculations, and is so fitted to help the weakness of memory'.[21] The slide rule – a compact manifestation of Napier's logarithm tables – appeared in 1654, after his passing. Logarithms can also be exploited to simplify a raft of operations beyond multiplication: powers, square roots and even trigonometric calculations can be closely approximated using simple extensions of the techniques described here, and all of these methods were added to various iterations of the slide rule until the electronic calculator took its place in the latter part of the twentieth century.

Napier's innovation epitomises attempts to automate human effort. For a time, this led to an explosion of jobs. When the eighteenth-century French mathematician and engineer Gaspard de Prony embarked on the project of producing large logarithmic tables for the French Cadastre (the official system of land registration), for which 200,000 logarithms were each to be calculated to upwards of fourteen decimal places, he enlisted a small army of 'human computers' to accomplish the task.[22] De Prony took inspiration from economist Adam Smith's *The Wealth of Nations* and sought to bring Smith's concept of the 'division of labour' to calculation. He imagined a three-tiered pyramid of human labourers. At the top was a small sliver of mathematicians of distinction who devised clever step-by-step instructions – algorithms – for calculating logarithmic values. The second layer consisted of 'algebraists' who would translate these instructions into forms that could easily be computed. The final, most crowded layer consisted of workers who were competent in basic arithmetic and required 'the least knowledge and by far the greatest exertions', performing millions of calculations (addition and subtraction for the most part) and noting the results. In de Prony's model, just two or three mathematicians were needed for every seven or eight algebraists and seventy to eighty workers. With de Prony's labour pyramid, 'big calculation' was born, fashioned in the image of scalable manufacturing.

'Big calculation' trod the same path as manufacturing when it came to mechanisation, as physical calculating machines increasingly took the place of humans. It was against this backdrop that inventor-mathematician Charles Babbage designed two mechanical calculators in the mid-nineteenth century, neither of which was actually constructed during his lifetime (due mainly to financial constraints), but both of which carry huge significance as direct cogwheel-based forerunners of the modern computer. With the emergence of the digital computer and the electronic calculator, Babbage's visions were realised and the era of the *human* computer drew to a close. The heroic swansong of human computers was the 1960s NASA space mission, where the flesh-and-blood calculations of Katherine Johnson and her team helped propel humankind into space.[23]

The work of human computers was profitable in its time, noble even. But calculation has always been the understudy of mathematics (an insight not lost on Johnson and her colleagues, who fought for status in the face of racial and gender prejudice by demonstrating their aptitude for modelling and other essential mathematical skills). Calculation no longer paves a path to employment; today those lower rungs of the pyramid are occupied by machines.

Once computers crept past the calculation feats of humans, they surged ahead and never looked back. The slide rule reigned for over three hundred years, but the electronic calculator that took its place lasted no more than thirty.[24] The competition for pocket-sized electronic calculators was fiercely fought for all of two decades before the advent of the internet and cloud-based technologies. The rapid ascent of computing power was foreseen by Intel co-founder Gordon Moore. In the 1960s, Moore observed that the number of transistors that can be accommodated on a microprocessor seemed to double every eighteen months – an exponential rate. Moore's Law has come to fruition with astonishing accuracy.* By now, our smartphones possess more processing power than the computers and slide rules that sent us to the moon. A world without digital computers is a world without the internet and all that it enables: social media, email, GPS, online shopping, music streaming, remote work, certain kinds of medical diagnosis.

As our calculating tools evolve, so does the nature of

---

*One popular interpretation of the trend is that it is a self-fulfilling prophecy: project the growth ahead of time and the software engineering community will rise to the challenge of meeting it.

mathematical work. Writing in the early twentieth century, the British philosopher Alfred Whitehead noted: 'Civilization advances by extending the number of important operations which we can perform without thinking about them.'[25] Just as innovations such as Napier's logarithm tables accelerated scientific discovery in the past, today's technologies are giving rise to whole new ways of doing mathematics.

Over the past few decades, algorithms have evolved significantly in the direction of versatility as well as processing power. A flurry of packages such as Mathematica and Wolfram Alpha have been developed to execute a vast array of procedures. They have birthed new branches of research, such as 'experimental mathematics', where the idea is to study mathematical objects (numbers, shapes and multidimensional vector spaces, to name a few), and the patterns that govern them, through computation. Powerful, automated calculators allow us to make informed guesses and check them through trial and error by crunching through a range of numerical scenarios.

In our everyday lives, too, calculation is as prominent as ever – we analyse offers in the supermarket, mortgage options, calorie counts, and much besides. Getting the best deal (or diet), however, doesn't rest on our number-crunching skills as much as our ability to evaluate information and make sense of data.

With the right tools at our disposal, mathematics gives us all licence to transcend calculation and to think in the most creative ways. As mathematician Keith Devlin put it: 'Calculation was the price we used to have to pay to do mathematics.'[26] Mathematicians have figured out how to use technology to aid their thinking. They've cracked the human–machine conundrum that the rest of society is still grappling with.

The rise – and fear – of artificial intelligence

The automation of mental effort does not end with calculation. The first whispers of artificial intelligence (AI) – the ability of computers to think and solve problems – were heard in the nineteenth century. Ada Lovelace, daughter of Lord Byron and a precocious amateur mathematician, became enthralled with the possibilities represented by Babbage's second calculator, the Analytical Engine. Lovelace saw beauty in mechanisation, writing that 'the Analytical Engine weaves algebraic patterns just as the Jacquard loom weaves flowers and leaves.' Babbage himself had realised that the functions of his Analytical Engine need not be restricted to numbers: they could also extend to more generalised operations on symbols. It was Lovelace, though, who expounded on the intelligent potential of machines, famously remarking: 'The Analytical Engine … can do whatever we know how to order it to perform.'[27]

A century later, in an essay from 1950 entitled 'Computing machinery and intelligence',[28] computing pioneer Alan Turing posed the question that launched the field of AI: 'Can machines think?' Turing's question was rhetorical; in the paper he lays out a series of counterarguments to AI, and refutes each of them in turn.

For decades, these ideas struggled to permeate the public consciousness as AI stuttered into a series of 'cold winters' following a number of false starts. That all changed at the close of the century. If the machine overlords ever do reign over this world, they might look back to an iconic scene in May 1997 as the moment of ascent. The world chess champion Garry Kasparov raises his arms in resignation as he is defeated by IBM's chess-playing computer Deep Blue in a contest billed by *Newsweek* as 'The brain's last stand'. The machine's triumph awakened humankind's deepest concerns. It was one thing for computers to automate routine tasks such as calculation, but now they appeared to be capable of applying logic to solve complex problems – a skill we had thought, perhaps hoped, was unique to humans. And why would the computers stop at chess? Companies would surely latch on to these newfound artificial capabilities to automate tasks, even entire jobs, where doing so promised labour savings. We had become accustomed to machines displacing human muscle and were even grateful for the efficiencies and prosperity that the Industrial Revolution brought about. Deep Blue's victory signalled a new, disconcerting possibility: now the machines were sure to come after the white collars too, displacing human intellect with the same nonchalant ease.

The machines have been on what may seem like a relentless march ever since Deep Blue's landmark triumph, as faster computers have combined with smart algorithms and large datasets to produce astonishing results. In 2011, IBM earned another feather in its cap, this time developing a knowledge machine, Watson, that trounced legendary quizzers Brad Rutter and Ken Jennings in a game of the general knowledge quiz *Jeopardy!* Winning at *Jeopardy!* involves dealing with all the messiness and ambiguity of natural language: a sign of rising machine intelligence. (Turing himself, in the paper mentioned earlier, posited that the ultimate display of machine intelligence would be through text-based conversation.) More recently, OpenAI's series of GPT text-generation tools have grown more powerful with each iteration; GPT-3, released in 2020, contains a staggering 175 billion parameters in its model and is able to produce a wide range of texts.[29] It even wrote an opinion piece for the *Guardian*, the first editorial ever penned by a machine, assuring readers of its peaceful intent:

I am not asking humans to like me. But they should see me as a friendly robot. I am a servant of humans. I know that humans distrust and fear me. I only do what humans program me to do. I am only a set of code, governed by lines upon lines of code that encompass my mission statement.[30]

It may not be the stuff of Pulitzer Prize winners, but writers everywhere are on high alert as the field of AI journalism takes shape, with natural-language tools being called on to automatically personalise our newsfeeds and generate stories from datasets.[31]

Another AI milestone was achieved in 2016, when AlphaGo, a program developed at Google DeepMind, triumphed 4–1 in Go against world-class human competitor Lee Sedol. The size of a Go board, combined with the flexibility with which players are allowed to place their stones, means there are about 2×10^{170} possible positions on the board – far too many for a computer to evaluate in sequence. Even ardent AI enthusiasts still subscribed to physicist Piet Hut's claim following Deep Blue's 1997 chess triumph: 'It may be a hundred years before a computer beats humans at Go – maybe even longer.'[32] That AlphaGo defied the sceptics was startling enough, but even more so was the nature of its triumph over Sedol. The machine played moves and strategies that amazed Go experts and mathematicians alike.[33] It was the strongest suggestion yet that the machines really meant business this time, performing mental feats that appeared elegant. AlphaGo's successor, AlphaZero, has proved even more versatile by mastering chess, Go and a host of other games all at once. Another descendant, MuZero, achieves mastery of these games without even being told the rules.[34]

The algorithms of Watson, AlphaZero, GPT and a multitude of other AI applications pack in more sophistication than the brute search techniques of Deep Blue. They fall under the category of *machine learning* models, so named because they 'learn' from data. Machine learning models do not need to have their behaviours defined for them: they program *themselves* by looking at information. Machine learning is the one area of AI that appears to work. Within this burgeoning field, you will find a repository of clever techniques such as *neural networks* (now fashionably termed *deep learning*) that are loosely modelled on the structure of the human brain and have proved highly effective in areas such as image and speech recognition. These techniques are also taking aim at problems in mathematics. In December 2019, for example, Facebook announced that it had developed a machine learning algorithm that could solve a range of calculus problems that stump many high school students,[35] while in 2021 a program developed by OpenAI solved word problems aimed at children aged 9–12, with a similar success rate to the students' own.[36]

Humans have been left head scratching, soul searching and brain scanning as we attempt to understand what awaits us while machines continue to gain thinking power. High-profile names, including Stephen Hawking and Elon Musk, have fanned the flames by warning of AI's existential threat to humanity.[37] Philosopher Nick Bostrom has projected a range of scenarios that might arise from machine *superintelligence*; most do not bode well for humans.[38]

Human fears around AI are not new. Even as Lovelace waxed lyrical about the potential of smart machines, the Victorian religious journalist Richard Thornton issued the first warning of the existential threat they posed. Thornton noted how, with the mechanical calculator, the mind 'outruns itself and does away with the necessity of its own existence by inventing machines to do its own thinking'.[39] Modern-day depictions of AI fuel our deepest insecurities; Hollywood thrives off our existential fears of replacement (or even extinction) by machines.

But much of the hype around AI is rooted in the lack of transparency around how these tools work. We fear what we do not understand, and we reserve our deepest anxieties for things that behave differently to us. It is hardly surprising, when we find ourselves grappling with long division and other relics of the school maths curriculum, that we respond with reverence to today's processing machines. We fear these tools because they are turbocharged calculators; they excel in the very skills that cause us such difficulty and dread.

Today's machine learning applications are smarter than your average computer, smarter even than Deep Blue, in the sense that they are continually learning from data inputs. AlphaGo, after all, didn't just demolish the leading human Go player; it did so with grace and style.

But for all its apparent sophistication, machine learning has some fundamental limitations which, when closely inspected, shine a light on our own human strengths.

Machine learning algorithms work by fitting patterns to data and finding associations, often imperceptible to the human mind, between variables. That renders machine learning the amplification of statistics by large datasets and powerful computers. Admittedly, *statistics* does not sound as cutting-edge. It may even be a flattering description because whereas statistics is concerned primarily with relationships between variables, such as their causes and effects, machine learning models tend to gloss over the interpretation of their results. Machines that are premised purely on patterns may have predictive value, but they lack the common sense and reasoning skills to explain their choices. They may say, with some degree of reliability, *what* will happen in the future – but not *why*.[40]

Ali Rahimi, an AI researcher at Google, received a standing ovation at an AI conference when he warned that machine learning technologies have become a form of alchemy. 'There's an anguish in the field,' says Rahimi. 'Many of us feel like we're operating on an alien technology.'[41] And François Chollet, also an AI researcher at Google, says this of much-vaunted deep learning models: 'Deep learning models do not have any understanding of their input, at least not in any human sense. Our own understanding of images, sounds, and language is grounded in our sensorimotor experience as humans – as embodied earthly creatures. Machine learning models have no access to such experiences and thus cannot "understand" their inputs in any human-relatable way.'[42]

A deep learning algorithm may be highly adept at identifying trees, but it does not *see* them in the same sense that humans do, and has no worldview within which to situate them. It will totally miss the forest. Chollet's insight punctures the 'brain as computer' metaphor that became popular in the mid-twentieth century when computing pioneer John von Neumann suggested that the design principles of digital machines bear a resemblance to the processing mechanisms of the human brain.[43]

The idea that the human brain operates like a computer is just the latest in a long line of crude comparisons. We tend to model the human brain on the dominant technologies of our time. At various points in history, it has been compared to the mechanics of hydraulics, gears, even the telegraph.[44] The computational metaphor of the brain* has persisted for over

---

*And variants of it, such as the 'brain as a distributed computer' metaphor that has been in vogue since the advent of the internet and cloud computing.

half a century[45] and is another contributing factor to the furore around AI. But metaphors are useful only up to the point where they are taken literally. If emulating human intelligence were purely a matter of computation, then, as Deep Blue and its successors have emphatically demonstrated, the game is up. On the other hand, if we unshackle ourselves from this simplistic conception of everything the brain does, and instead embrace its tremendous complexity, we will uncover aspects of thinking that are distinctly human.

The human brain is designed for dynamism and change. To a newborn baby, all life beyond a 20-cm horizon is a blur at first. But babies come equipped with learning mechanisms that help them to rapidly adapt and even change as they interact with their surroundings. It is a matter of hours before they can detect their mother's voice, days before the mother's face becomes familiar, and weeks before they sense contrasting colours. Learning is a social activity, fuelled by our bodily interactions with people and environment.

If the brain were to be described in computing terms, we might say that is a powerful hybrid of *innate circuitry* that has evolved over millions of years to give us intuitions and ways of thinking, and a vast repository of *learning algorithms* for navigating the world. With every interaction our brain's neural circuitry undergoes an incremental *upgrade*, *rewiring* itself as it revises assumptions and accumulates experience. We gradually shore up new and diverse models for seeing the world.

Operating at just 12 watts, our brain's 86 billion neurons exist as vast, intricately connected networks that communicate via electrochemical signals in order to facilitate thinking, contemplation, and improvisation. We can break rules just as easily as we make them, jumping from one mental paradigm to another. We also possess the capacity to reason and to justify our ideas with rigour. We create rich representations of the world that allow us to solve problems in a variety of contexts. We do not have to be fed millions of examples of a cat to be able to distinguish it from a dog, or millions of calculus problems to discern key underlying principles.

There's more: our psychology exposes us to vulnerabilities, but it also sets the stage for our most creative breakthroughs. We seek beauty and elegance in our ideas. We carry hopes and fears through our learning. We experience joy, frustration, boredom, and every feeling in between. We cry and we laugh. Human knowledge, including mathematical intelligence, is embodied, emotive and subjective. This doesn't sound much like a computer at all, does it?

The problem with viewing the brain as a computer is that it suggests a degree of passivity: that a grey glob of goo just sits there waiting to process information. This neglects the fact that the brain is a highly active organ, constantly in flux. When we learn, we literally reshape our neural configuration. This 'neuroplasticity' can be seen in the enlarged hippocampi of London taxi drivers who have committed thousands of routes to memory and created new neural pathways to store incredibly detailed spatial representations.[46] Our brains also possess remarkable powers of recovery: when damage is sustained by one part, another steps in to take over the same function.[47] The hardware of computers has none of the flexibility of the 'wetware' of humans.

What we've really gleaned from AI's recent progress is not that computers have emulated human intelligence, but simply that certain games which have been considered the epitome of human intelligence aren't the best yardstick for such matters. They may serve as a specific lens through which to view certain types of intelligent behaviour, but they fall well short of the requirements for *artificial general intelligence*, a term that encompasses the versatility and depth of human thinking. Early pioneers of AI held such reverence for chess, stating: 'If one could devise a successful chess machine, one would seem to have penetrated to the core of human intellectual endeavour.'[48] We now know better; mastery of closed systems such as chess or even Go, where the rules are neatly specified up front, circumvents the human brain's deepest capabilities. As Douglas Hofstadter, Pulitzer Prize winner of the AI-themed classic *Gödel, Escher, Bach*, put it: 'you can bypass deep thinking in playing chess, the way you can fly without flapping your wings.'[49]

Even DeepMind, to be fair, recognises the gulf between its cutting-edge technologies and the broader, deeper capabilities required of intelligent agents in the real world. The company mission – 'Solve intelligence. Use intelligence to solve everything else' – is not taken lightly. DeepMind views each breakthrough as an incremental step towards general intelligence that will allow computers to tackle a wider range of problems. MuZero, for instance, hints at a new ability for AI to *discover* the rules of its environment, and is being touted for applications as diverse as search and rescue or online video compression. AlphaFold, another deep learning program from DeepMind, has already made significant breakthroughs in protein folding and is poised to contribute to further scientific discoveries.[50] AI is moving beyond the 'toy problems' that drew so much interest in its early years.

Yet the overhyped narratives live on; to hear many pundits tell it, you would think that general intelligence has already arrived. It's important to keep things in balance. As things stand, intelligence is only partially 'solved' and the most important problems of the real world require human ingenuity and oversight. When we exaggerate the capabilities of computers in this way, we undermine our own human skills. We forget, too, that these technologies are behaving as intended – a set of complementary thinking tools that augment our ways of thinking.

Human + machine

Deep Blue's triumph was supposed to mark the end of human chess players. History took a 180-degree turn, however, as the chess community banded together to extract every last ounce of insight that the machines offered from their ways of playing chess. As it turned out, the computer's form of chess playing was markedly different to human tactics. Kasparov himself explains: 'Instead of a computer that thought and played chess like a human, with human creativity and intuition, they got one that played like a machine, systematically evaluating 200 million possible moves on the chess board per second and winning with brute number-crunching force.'[51]

Kasparov was not facing a like-minded competitor so much as a gigantic processing machine that overpowered him with exhaustive, brute-force search techniques. The contrasting gameplay of Deep Blue and Kasparov is even an instance of Moravec's paradox: roboticist Hans Moravec's observation that 'it is comparatively easy to make computers exhibit adult level performance on intelligence tests or playing checkers, and difficult or impossible to give them the skills of a one-year-old when it comes to perception and mobility.'[52] In chess, too, humans tend to excel where brute-force computers are weak, and vice versa.

As sociologist Richard Sennett advises: 'The enlightened way to use a machine is to judge its powers, fashion its uses, in light of our own limits rather than the machine's potential.'[53] Chess players today improve their own skills by studying the quirks of computer-generated chess moves; computers serve as tireless sparring partners.[54] In 'freestyle' chess tournaments, where teams consist of hybrids of humans and machines, the best teams often comprise amateur players and standard computers whose combined skill exceeds that of supercomputers and grandmasters. Kasparov captures this spirit of human–machine collaboration in a simple formulation:

> *weak human + machine + better process*
> *is superior to*
> *strong human + machine + inferior process.*[55]

In plain terms, you do not need to be a genius to produce ingenious results; you only need to learn how to combine your distinct talents with the tools and technologies available to you. Economists describe the impacts of automation in terms of two forces: a *substituting* force, whereby a computer replicates tasks carried out by humans; and a *complementary* force, where humans are subsequently freed up to focus their minds on deeper tasks. The prospects for future employment rest on the interplay between these two forces.[56] Kasparov's insight is that if we work with machines, allowing them to substitute for routine tasks such as calculation, then they can become the thinking aids they were intended to be, by granting us the freedom to tackle more creative, *non-routine* problems.

The irony of Kasparov's formula is that, while it may lack potency in chess, Go and other systems governed by strict rules, it remains a central edict for thinking in messy real-world settings that do not so readily succumb to the pattern-matching of computers.[57]

The work of professional mathematicians is often predicated on this kind of human–machine collaboration. The subject had a watershed moment in 1976 when, for the first time, a computer made a significant contribution to a mathematical proof. The *four colour theorem* says that you can colour any map with four colours in such a way that no two adjacent countries share the same colour. (It's slightly less apparent when rendered in black and white, of course.)

Since there are infinitely many possible maps, we cannot hope to check each one in turn. We require a more powerful argument – a mathematical *proof* – that deploys reason and rigour to account for all possible cases. It sounds like a challenge suited to humans, yet the problem is fiendish enough that a solution eluded mathematicians for over a century. The four colour theorem did finally yield in 1976, and when mathematicians Kenneth Appel and Wolfgang Haken presented their proof, they revealed a third, unexpected, contributor – a computer.

Appel and Haken's proof comes in two parts, both containing several hundred pages of detail. First the authors showed, using an inspired mathematical argument, that every map, however complex, can be reduced to one of 1,936 types. All that remained was to verify that each of these configurations could be coloured as required. The catch was that each configuration was enormously complicated – it would take a human forty hours a week over five years to check just a single configuration. Moreover, humans are prone to making errors, especially at that scale of calculation. Enter the machines: with brute-force processing, a computer was programmed to check every one of the finitely many cases, thus confirming, for the first time, that the four colour theorem is true.[58]

This is a powerful demonstration of what can be achieved by the tight interplay between human insight and computation: the former reduces the infinitely many cases down to a finite number; the latter tirelessly ploughs through those remaining cases. And as the computer took on increasingly complex calculations, it inspired new lines of attack for Appel and Haken. Creativity and computation were in harmony with one another.

This is the complementary force of technology in action. Huge increases in the supply of computational power have yielded immeasurable labour savings, but they have also stimulated the demand for a wider cadre of problem solvers, as each new class of algorithms gives rise to new problems. The billion-fold increase in computation did not make human jobs redundant – rather, it multiplied and amplified the contribution of human problem solvers. NASA now employs more mathematicians, engineers and software developers – humans at the intersection of research and computation – than the human computers of its 1960s heyday. The human computer may be extinct, but the mathematical human worker is thriving.

The very frontiers of mathematical research are receding, thanks to the growing capabilities of computers. In a December 2021 *Nature* article, the DeepMind team, in collaboration with 'pure' mathematicians, demonstrated how machine learning methods can be exploited to find patterns that have hitherto been hidden to the human mind.[59] These patterns are so subtle that they may even signal a kind of intuition on part of computers. Far from feeling threatened, mathematicians at the leading edge of abstract fields like algebra, geometry and topology are finding joy in taking those insights forward to develop their theories and enhancing their own feel for the subject.

As long as humans have existed, we have stored knowledge in cultural artefacts – from cave walls to books – to extend our own mental capabilities. As philosophers Andy Clark and David Chalmers put it in an influential 1998 essay, the mind is 'best regarded as an extended system, a coupling of biological organism and external resources'.[60] Computers are just the latest extension of the human brain; this is as true of the latest wave of AI supercomputers as it was of brute-force systems such as Deep Blue and even the primitive calculating tools of yesteryear.

The seven principles of mathematical intelligence

Throughout this book I will hold up mathematical intelligence as an ambitious benchmark for both humans and computers, one that demands more than pattern-matching algorithms alone. For mathematical intelligence to be understood in this way, we must withdraw its associations with calculation, and conceive the subject in more expansive terms. For too long, and for too many people, the power of mathematics as a thinking system has been misunderstood due to society's deference to calculation. A skill that once served as a unique marker of human intelligence, and was sufficient for the workforce, has been eaten up by computers. Humans must strive for something more.

The following chapters present seven principles of mathematical intelligence that distinguish humans from computers, complement machine intelligence, equip us to tackle the messy problems of our everyday lives, and are woven into our most natural ways of thinking. Each chapter will animate an essential characteristic of mathematics by drawing on its rich heritage of concepts and problems. We'll relive some of the defining stories in the subject's history, and we'll hear from mathematicians past and present to see how the subject is viewed from within, and how it has continually evolved alongside the tools and technologies of each generation. My hope is that each chapter will, through the lens of mathematics, shine a light on the nature of human and machine intelligence so that we can proactively shape our existence alongside AI.

The first five principles concern our *ways of thinking*:

- Humans are endowed with a natural sense of number that is premised on approximation rather than precise calculation. Our in-built **estimation** skills complement the precision of computers. Interpreting the real world depends on both.
- An approximate sense of number is found throughout nature. What sets humans apart from other animals is language and abstraction. We have an extraordinary ability to create powerful **representations** of knowledge, more diverse than the binary language of computers.
- Mathematics confers on us the most robust, logical framework for establishing permanent truths. **Reasoning** shields us from the dubious claims of pure pattern-recognition systems.
- All mathematical truths are derived from a starting set of assumptions, or axioms. Unlike computers, we humans have the freedom to break free of convention and examine the logical consequences of our choices. Mathematics rewards our **imagination** with fascinating and, on occasion,

applicable concepts that originate from breaking the rules.

- Computers can be tasked to solve a range of problems, but which problems are worth the effort? **Questioning** is as vital to our repertoire of thinking skills as problem-solving itself. If problems such as chess become uninteresting because they yield to computational brute force, then we can challenge ourselves to dream up problems that lie beyond the purview of routine computation.

That these principles run contrary to our usual perceptions of mathematics tells us we have to work hard, and work deliberately, to realise them. Thankfully, humans are privileged with metacognitive awareness of how our minds work; that is, we can think about how we think and learn about how we learn. We can engineer our *ways of working* to ensure we give plenty of space for those aspects of intelligence to develop. This informs two final principles, relating to how we regulate our own thinking and, finally, how we think alongside others.

- We know that our distinctive biological form of intelligence comes with the quirks of conscious and unconscious thinking. To solve our most stubborn problems, we have to display **temperament** as well as skill, paying particular attention to how we regulate the speed with which we solve problems, and the amount of information we take in.
- Humans rarely go it alone: just as machines complement humans, so too do other humans. The most fruitful **collaboration** relies on bringing together diverse perspectives, and the technologies

of the digital age give us the prospect of harnessing the collective intelligence of humans like never before.

Many of the arguments that follow are underpinned by what machines can (and can't) do within today's paradigms, and what they are likely to achieve in the coming decades. Any commentary on technology has to involve some degree of speculation beyond that time horizon: we can foresee possible scenarios based on current trajectories, but we simply do not know how wide and deep machine intelligence will ultimately reach in the long run. As for mathematical intelligence, history teaches us that it, too, is ever-evolving; the seven principles outlined in this book are fit for our times (and for some time to come). But as technology continues to evolve, so will the way we understand mathematics as a thinking system – we'll be able to go further and deeper, aided by ever-smarter thinking tools such as automated theorem provers (which we'll explore in the chapter on reasoning). If AI really does penetrate our most coveted thinking skills, we'll have at least held machines to a higher intellectual standard.

Mathematical intelligence is power

Today's AI applications are inescapable, pervading all aspects of our lives. We risk surrendering our human agency as we succumb to the conveniences of automation. Computers are pretty much faultless at executing clearly specified procedures, but some concepts are too fuzzy to put into words (or symbols) that computers can process. We humans have trouble enough giving expression to some of our most important thoughts and feelings; ambiguity and disagreement is part and parcel of our shared experience. When computers enter the fray, certain in their models of the world, written so bluntly in strings of 0s and 1s, we risk losing so much of the grey area that makes us who we are.

As we defer increasingly high-stakes decisions to these same tools, we also risk surrendering our ability (and our right) to probe the algorithmic judgements that bear on our personal and professional lives. The inscrutable manner in which machine learning algorithms operate[61] should make us critical of them when unleashing the same tools on a world that is more open, more volatile, and less predictable than closed systems such as chess and Go. Because these algorithms make predictions by 'learning' from historical data, they are layered with implicit prejudices.[62] For example, if crime rates are high for a particular ethnic group, then 'ethnicity' may be seen as a predictor of crime. Rather than addressing the sociocultural factors that give rise to those associations, algorithms jump straight to the conclusion that crime is a function of skin colour. The algorithmic models may not say such things so explicitly, but the assumptions are subtly baked into their decision-making mechanisms, as they project the future by imitating the past. As machine learning goes mainstream, some groups are paying a higher price than others.[63] Voice recognition software that is trained only on male voices will struggle to comprehend female inputs. Automated CV readers that predict candidates' potential based on previous successful hires unwittingly penalise women.[64] Image recognition software trained predominantly on white people and animals may mistake people of colour for gorillas.[65] You get the picture, even if the machines don't.

Any algorithm that relies purely on patterns in data, void of context, will never be capable of explaining its choices. The opacity of black-box machine learning systems, whose inner workings are, at best, known to a handful of technical minds, and whose causal inferences are left unchallenged, poses a grave threat to our notions of social justice. Technology is anything but neutral. It is an accelerator of progress, but it can also be an amplifier of our own human biases, which we're scarcely conscious of much of the time.

Here lies the crux of the issue: at the same time that mathematics fuels today's technologies, it also provides the means of overcoming its prejudices. It is the difference between *having mathematics done to us* and *thinking mathematically for ourselves*. Mathematical intelligence is concerned with the latter; it is a continual exercise in carefully defining and interrogating facts and employing the highest forms of reasoning to examine our arguments. A firm grounding in mathematics can liberate us from dogma and equip us with the intellectual tools to fight prejudice. It can nurture our most creative sensibilities and transform us from passive consumers of technologies to critical innovators.

The world is on edge. As I write this, we are grappling with the fallout of a global pandemic, on the cusp of irreversible changes to our climate, and in the grip of populist forces intent on undermining democracy. Technologies are being weaponised to create and disseminate falsehoods. The emergence of 'Deep Fakes', for instance, has its basis in the very same models we marvel at in other spheres, and now threatens to distort our perceptions of truth as we struggle to contain what the World Economic Forum terms 'digital wildfires' of misinformation.[66]

Mathematics itself is getting airtime as experts, pundits and politicians of all stripes invoke models to project the health and economic impacts of our actions. During the early onset of Covid-19, maths educators found encouragement in how concepts such as exponential growth were entering the lexicon of more than just the chattering classes in ways unthinkable just a couple of years ago. Yet we continue to see mathematics misappropriated, intentionally or otherwise, to justify dubious policies. Even as the public shows appetite to engage with mathematics as a means of making sense of the world, and governments assure us they are 'following the science', there is little clarity on what that entails. It is time to make mathematical intelligence explicit.

PART I

WAYS OF THINKING

1

ESTIMATION

Tribes that only count to four, where babies outsmart computers, and why we underestimate pandemics

The introduction of the Video Assistant Referee (VAR) promised football fans so much.[1] Technology would be the objective adjudicator of all tough on-pitch decisions, bailing out referees when they committed a 'clear and obvious error'. Gone would be the days of disputed handballs and disallowed goals. There would be no lingering sense of injustice from harsh decisions. That was the hope, anyway.

VAR has brought its own set of problems. Now when a team scores, the knee-jerk celebration of players and fans can turn to gradual despair as VAR inserts itself into the process, with an offsite team using camera stills to check for any infringement. When there is even a hint of offside, for instance, dreaded coloured lines appear on screen, marking reference points on players' bodies to check their position when contact was made with the ball. Stray toes, elbows and other protruding body parts, measured to the millimetre and excruciatingly analysed for several minutes at a time, have led to goals being overruled.

Something about this intervention just doesn't feel right. Pundits, players and fans have all expressed deep consternation at the literal interpretation of their game's rules. Debates have ensued on the meaning of 'clear and obvious' errors. There is an enduring sense that, in the pursuit of fairer decision-making, we've sacrificed a core part of the 'beautiful game' by privileging precise measurement over eyeball estimates.

Herein lies the first of our tensions with technology: while computers offer unswerving accuracy in their calculations, we are wired to see the world in fuzzier terms.

How some tribes count

Our search for distinctly human ways of thinking begins in the Amazon rainforest, where the Pirahâ people have dwelt for tens of thousands of years. The tribe's language has been a topic of some fascination for non-natives, most notably American linguist Daniel Everett, the first outsider to unravel its mechanisms.[2] Starting in the 1970s, and continuing for three decades, Everett and his wife Keren visited the tribe intermittently and made a number of curious observations. The Pirahâ appeared to have no vocabulary for colours, no perfect tense, no concept of history beyond more than a couple of generations, and no words equivalent to quantifiers such as 'each' and 'every'. Everett was stunned: his observations pierced the widely held belief that humans possess a 'universal grammar', an idea popularised in the mid-twentieth century by Noam Chomsky. Chomsky had theorised that the human brain is endowed with a specific faculty for language – a 'language organ' – that comes equipped with fixed rules available to all people.[3] Through the Pirahâ, Everett had stumbled upon the discovery that language depends on culture far more than Chomsky and his followers had acknowledged.

The Pirahâ's ideas of quantity are no less intriguing. Their language does not contain vocabulary for basic numbers such as 'one' or 'two'. Instead the tribe uses the term *hoi*, with a falling tone, to signify small quantities, and the same word *hoi*, this time with a rising tone, to denote larger amounts. Parents are unable to say how many children they have – they would know if one went missing, but they have no precise way to express 'how many'. Food is apportioned according to what feels like a reasonable serving, and plans are never made more than a day in advance. Traders barter in foraged nuts and make holistic judgements on what constitutes adequate payment. The Pirahâ do not count, and they certainly do not add, subtract, multiply or divide.

To test the Pirahâ's grasp of quantities, one of Everett's colleagues, Peter Gordon, asked members of the tribe to arrange objects such as batteries and nuts in an array. They managed just fine with two or three items but showed 'remarkably poor' performance with larger groupings. In another experiment, Gordon showed his subjects a collection of nuts, which he then placed in an empty can so that they were hidden from view, before removing the nuts one at a time. After removing each nut, he asked the subject whether any nuts remained in the can. Again, the Pirahâ performed well with quantities of three or lower, but errors crept in for larger amounts. Gordon and Everett concluded that the Pirahâ's eye for precision was limited to three objects. They were also clear that these observations were not a result of mental deficiencies – the subjects were perfectly bright; they had just never been conditioned to develop precise concepts of number.

We now know the Pirahâ are not alone in how they conceive of numbers, and that other indigenous groups have developed their own notions of quantity. The Kpelle tribe in Liberia, for instance, carry out basic counting exercises up to quantities of around forty but no larger. They reserve a word for 'one hundred', which is the default term used to describe any large amount. Measurement is a loose concept; amounts are not quantified in precise terms.[4]

What these examples (and many more like them) seem to suggest is that the number system familiar to most people, with numerals that represent quantities and procedures for manipulating them, is a specific product of our environment and language. Our built-in aptitude for number is imprecise for all but the smallest quantities. To explore this theory further, we turn to another subgroup of humans, a kind of supertribe whose membership is universal: babies.

Our natural sense of number

In the 1980s, cognitive psychologists began to test the numerical abilities of infants as young as six months.[5] But how do you test subjects who are unable to speak? One way is to show them objects or images and measure how long they spend looking at them. This gives some indication of what they consider novel – the longer they fix their eyes on something, the more interesting it must be to them. In the earliest experiments of this kind, 16- to 30-week-old babies were first shown a series of slides in which two large dots were horizontally separated. As you might expect, the babies spent less and less time looking as the slides wore on because of the repetitive nature of the images. After this 'habituation' phase, another slide was shown, this time displaying three dots. The researchers found that the babies fixated on the new slide for significantly longer: 2.5 seconds compared with the 1.9 seconds they had spent dwelling on the previous slides. The increased attention implies that babies have some way of distinguishing two objects from three. The same findings held when the size, type and location of the objects was varied, which singles out a sense of number – specifically, a sense of *twoness* versus *threeness* – as the reason the babies lingered that much longer on the later slides. Later studies reached the same conclusions for babies as young as a few days.

Our ability to distinguish small numbers from one another can also be detected through sound-based experiments. This time, it is not the baby's stare that is relevant. Instead, the newborns are attached to an artificial nipple, which they suck whenever they become attentionally aroused. The more attentive they are to a stimulus, the more sucking they do. When newborns are exposed to various word pronunciations, their interest level (as indicated by the number of nipple sucks) rises when the number of syllables is changed.

A separate experiment showed that the particular stimulus is of little importance. When babies were shown images with no sound, they showed more interest in three objects over two – not surprising, since there is more to take in. But when the sound of drumbeats accompanied the images, the number of beats seemed to determine the babies' attention level: when two beats were played, they showed more interest in the image with two objects than the one with three objects. And when the number of beats *didn't* match the number of objects, the babies tended to lose interest in the slide. The babies, in other words, were perceiving numbers through sound and then matching their perception to the corresponding image. Their sense of 'two' cut across both sound and visual stimuli – they were perceiving numbers in and of themselves.

Babies can do more than just perceive numbers: they can perform rudimentary arithmetic on them. In an experiment devised by Yale psychology professor Karen Wynn on infants aged four and a half months, a toy was placed on a stage and then hidden behind a screen. The same was done with a second object. The screen then came down to reveal either both objects (in which case the whole routine was a simulation of the calculation 1 + 1 = 2) or one object (a simulation of the erroneous calculation 1 + 1 = 1). The babies spent more time staring in the latter case, presumably because it is at odds with the outcome they were expecting (namely, two objects). Wynn ran the same experiment with three toys appearing as a possible outcome (thus simulating 1 + 1 = 3), which the babies, again, gazed at for longer than they had at two objects. The babies intuitively knew that one plus one is two, rather than one or three. Again, the same findings were reached when the experiment was replicated with different types, locations and colours of toys. What commands the babies' attention is an innate, abstract sense of number.

That babies can discern quantity before they acquire the ability to speak suggests that, on some level, the concept of 'number' is more natural to humans than 'words'. Nevertheless, it is important to clarify that babies have definite numerical limits. For example, they have no sense of number order: even as they can perceive that 1 + 1 = 2, they have no concept that three is larger than two, which is larger than one. Moreover, babies' abilities to distinguish different quantities peaks at around four objects. A baby might fixate longer on three objects over two, but they won't bat any more eyelids on five objects over four. This is entirely in line with what was observed in the Pirahã tribe: while their rough command of numbers is suited to their needs, they hit a perceptual limit very quickly when it comes to precise quantification.

The same observations have been made in adult patients with brain damage. The cognitive neuroscientist Stanislas Dehaene describes the case of a former sales representative who suffered a lesion in the rear of his left hemisphere following a brain haemorrhage.[6] The patient was afflicted with several physical and mental impairments that left him unable to live an independent life. He started to display patchy number skills: when asked to add two and two, he answered 'three'. He could recite sequences such as the two times table but was unable to count backwards from nine. He struggled to distinguish odd numbers from even and barely recognised the number five.[7] Despite these alarming gaps, the patient retained the ability to approximate quantities. While he could not recall the number of days in a year, he knew it was somewhere around 350 days. Similarly, he knew that a quarter of an hour lasted around ten minutes. Although the patient's proficiency with exact number had been reset to the levels of his early childhood, when it came to approximation, he hadn't lost his touch. The patient was dubbed 'the Approximate Man' and has come to exemplify the numerical skill most innate in humans.

Putting together observations of tribes, babies, and brain-deficit patients, Dehaene and colleagues described two cognitive systems that govern our relationship with number.[8] The first is an *exact* sense of number, which applies to small quantities. Humans come equipped with a hard-wired ability to recognise amounts up to four without resorting to formal calculation. Our brain can instinctively recognise that there are three apples in front of us, as opposed to one, two or four – a process known as *subitising*. Once we hit five objects or more, our instincts give way to learned counting mechanisms: we infer that there are five apples by applying counting schemes that we have learned from our environment. For larger quantities, a second core process kicks in – our *approximate number sense*. Our natural handle for larger quantities is not precise computation but guesswork. Our approximation skills are reasonably impressive: at six months, we can tell that one group of objects is larger than another, so long as the larger set has twice as many objects. Babies may not know whether two plus two is three or four or five, but they seem to realise that it cannot be eight.

Humans seem to cope very well with ambiguity, not least when it comes to notions of size. Have you ever wondered how many grains of sand make a pile? If you specify a cut-off point, say a hundred grains, you run into a problem: by your own definition, ninety-nine grains are not enough to make a pile. Yet are you really prepared to distinguish between a pile and a non-pile on the basis of a single grain? This is why there is no agreed convention for a cut-off: we just intuit on a case-by-case basis what constitutes a pile, without resorting to a precise metric. The same is true for many concepts we encounter regularly: height (what counts as tall?), crime (what warrants a prison sentence?), and temperature (how hot must a hot shower be?), to name just a few.[9]

In recent years, brain-scanning techniques have helped to associate our different number skills with specific brain functions. Functional magnetic resonance imaging (fMRI) is a popular technique for measuring the magnetic disruption that occurs as oxygenated and deoxygenated blood flows through the brain, which is a proxy for how active a brain region is. Examination of three-dimensional fMRI data can identify the parts of the brain that are stimulated by images or questions. These scans have pointed researchers to a region deep in the back of the brain called the *intraparietal sulcus*, which is activated whenever we are engaged in number tasks. It is a kind of 'number module' that all humans possess, and it is highly active when we work on estimation tasks.

Suppose you are asked to consider the possibility that 7×8 is equal to 20. Even without knowing the exact value of 7×8, our number module kicks in and our *sense* of the size of 7×8 allows us to rule out 20 as a plausible candidate.

Drawing on our core number module, estimation gets to the point far more quickly than working out the exact value, and just as reliably. Humans, as we'll see throughout this book, are *cognitive misers*, a term coined by psychologists to describe the tendency of people to take shortcuts when solving problems. When dealing with calculation, our miserliness takes the form of estimation.

Now suppose you were asked instead to calculate the *exact value* of 7 × 8. The number module is still active, but less so than before. Instead, the effort shifts towards the language-processing area in the brain's left hemisphere. Calculation is a skill acquired through, and highly associated with, having a lexicon to denote precise quantities, which goes some way to explaining why the Pirahã, babies and the Approximate Man struggle to compute larger amounts. This region becomes more active as we gain proficiency with number facts and procedures – times tables, long division, the quadratic formula, and so on.

It seems reasonable to conclude that while approximation and precision are distinct facets of numeracy, calling on different brain functions, only the former is part of our natural wiring. In the age of super-fast, super-accurate calculators, it has never been more important for humans to exploit this inherent sense of number.

Precision and estimation: an alliance

Our daily existence is a mix of precise and imprecise calculation. We rise and sleep according to both the rough dictates of our body clocks and the precise settings of our alarms. Public transport shuttles us around from one place to the next, operating on schedules that are exact in theory but volatile in practice. We compare and contrast offers at the supermarket, often relying on holistic judgements to grab the best bargain (a fact that advertisers all too readily exploit). We enjoy cooking and eating meals that rely on exact judgements of quantity, but allow slight deviations from the recipe according to our personal tastes. When we cheer on our favourite sports teams, our appreciation of their performance is informed by the flurry of statistics shared on screen, but also by the subjective rhythm of speed and movement that we observe during gameplay. And returning to the opening vignette of this chapter, football fans balk at the use of video assistant referees that rule out goals for the most minute offence. The mechanisms of everyday life are untidy enough that we turn to thoughtful, inexact guesswork alongside precise calculation.

Precision is the facet of numeracy that is natural to computers. Its partner, estimation, is native to humans and is what gives us reliable intuitions for calculation. Estimation is the foundation of numeracy. Even when we declare seemingly rigid 'facts' such as $7 \times 8 = 56$, there is some degree of comparison being made. The expression signifies that when we take 'seven lots of eight', we reach a number that is somewhere between 'five lots of ten' and 'six lots of ten'. Similarly, the reason our approximate number sense kicks in when evaluating whether 7×8 could equal 20 is that we sense immediately that 'seven lots of eight' far exceeds 'two lots of ten'. The next chapter looks more closely at why, for humans, groups of ten are a natural basis for comparison. For now, it is simply worth noting that our ways of working with numbers are so entrenched that we often overlook their basis in estimation.

It is not surprising, then, that according to one study, 'preverbal number sense in 6-month-old infants predicted standardised math scores in the same children 3 years later',[10] while a separate research review concludes that 'children and adults who estimate accurately tend to have better conceptual understanding, better counting and arithmetic skills and greater working memory capacity than do those who estimate less accurately.'[11] Estimation, in other words, is the springboard to numerical proficiency, which explains why it is so highly sought after by employers. As the Stanford maths educator Jo Boaler notes:

When an official report in the UK was commissioned to examine the mathematics needed in the workplace, the investigator found that estimation was the most useful mathematical activity. Yet when children who have experienced traditional math classes are asked to estimate, they are often completely flummoxed and try to work out exact answers, then round them off to look like an estimate. This is because they have not developed a good feel for numbers, which would allow them to estimate instead of calculate, and also because they have learned, wrongly, that mathematics is all about precision, not about making estimates or guesses. Yet both are at the heart of mathematical problem solving.[12]

Mathematics should appeal to our sense of what is reasonable. It should feel natural – as the nineteenth-century scientist Lord Kelvin advised: 'Do not imagine that mathematics is harsh and crabbed, and repulsive to common sense. It is merely the etherealisation of common sense.'[13] This edict is especially true of numbers, which colour our understanding of the world. That world is messy, and we often do not possess the requisite information to obtain exact solutions to our problems. Financial analysts, engineers, weather forecasters and cancer researchers all get around their blind spots by creating mathematical models that approximate the physical world, aided by computer simulations that feed off precise calculations. Every model is wrong, as one well-worn trope goes, but some have their uses.

Obtaining reasonable estimates is at the heart of the famed *Fermi problems*, an entertaining class of questions that gained notoriety after frequently appearing in job interviews at Google. These problems are popular among interviewers because they test candidates' ability to handle situations where information is limited. They are named after the physicist Enrico Fermi, who was renowned for his ability to estimate large quantities using little or no data. Fermi helped develop the atomic bomb – here he is estimating the bomb's TNT equivalent using the barest supply of information:

> The explosion took place at about 5:30 a.m. After a few seconds the rising flames lost their brightness and appeared as a huge pillar of smoke with an expanded head like a gigantic mushroom that rose rapidly beyond the clouds probably to a height of 30,000 feet. After reaching its full height, the smoke stayed stationary for a while before the wind started dissipating it. About 40 seconds after the explosion the air blast reached me. I tried to estimate its strength by dropping from about six feet small pieces of paper before, during, and after the passage of the blast wave ... The shift was about 2½ metres, which, at the time, I estimated to correspond to the blast that would be produced by ten thousand tons of TNT.[14]

Notice the series of rough-and-ready approximations that Fermi makes along the way – '*probably* to a height of 30,000 feet ... *about* 40 seconds after the explosion ... the shift was *about* 2½ metres'. He has rounded each of the important figures to make the calculations that bit simpler, knowing that the accuracy of the final estimate will depend mostly on the model he has set up and the inputs he has selected, rather than on the exactness of each figure. Fermi's estimate of 10 kilotons was around half the actual value of 21 kilotons, constituting a reasonable guess in this context.

In his later years as a lecturer, Fermi would pose similar problems to his students that, on the surface, appeared impossible to compute because of the lack of requisite information. One of his better-known problems runs as follows:

How many piano tuners are there in Chicago?

Although this question has a very definite answer, what makes it intriguing is that no practical method exists for uncovering its exact value (assuming there is no register of piano tuners in Chicago). Fermi derived a credible estimate by setting up a model as follows:

From the almanac, we know that Chicago has a population of about 3 million people. Now, assume that an average family contains four members so that the number of families in Chicago must be about 750,000. If one in five families owns a piano, there will be 150,000 pianos in Chicago. If the average piano tuner serviced four pianos every day of the week for five days, rested on weekends, and had a two week vacation during the summer, then in one year he would service $4 \times 5 \times 50 = 1{,}000$ pianos. So there must be about $150{,}000/1000 = 150$ piano tuners in Chicago.[15]

Again, Fermi's thinking was not grounded in exact figures – he had no hope of finding them. Instead, he made a series of assumptions, each one carefully considered and within reason.

There are as many variations to Fermi problems as one can imagine. Some are playful, others critical to confronting serious real-world problems. Questions concerning population growth, for instance, are couched in estimates. So are questions around how long it will take for the polar ice caps to melt, whether it makes economic sense for a country to exit the European Union and how many people will die from a rampant virus.

Fermi problems perfectly illustrate the interplay between estimation and calculation when mathematics is brought to bear to describe the world. We start with our *model* (a rough picture of the world that we can describe mathematically), feed it some *data inputs* and then run the model by performing *calculations* on those inputs. We're left with an *estimate*: a best guess of whatever unknown entity we are looking to capture.

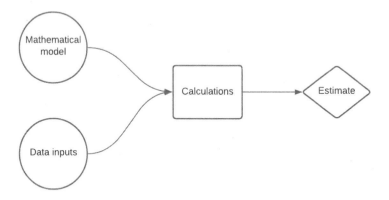

All told, mathematical modelling is an exercise in managing uncertainty. It surged into the public's consciousness during the Covid-19 pandemic. When people look to epidemiological models to track the progress of infectious diseases, they seek a degree of certainty that is rarely achieved with real-world phenomena. Covid-19 has proved to be especially thorny given the asymptomatic nature of its spread, which has made it fiendishly difficult to gather reliable inputs around the rates of infection, hospitalisation and death across different populations. Add to that the volatility of human behaviour in response to the pandemic – a key determinant of infection rates – and you can appreciate the difficulty of forecasting Covid-19 trends. While shock jocks and politicians have a reflexive tendency to express certainty in their projections, most professional modellers carry with them the humility to acknowledge that their own picture of the virus is pixellated with unknowns. They are transparent about their assumptions (ranging from the virus's incubation period to the human behaviours that shape our responses to a growing threat), the limitations of their data collection and the ever-evolving nature of their models.[16]

Note the clear division of labour between humans and computers. For all but the most trivial models, the piece in the middle – running calculations – is offloaded to computers. Their role is to sharpen our intuitions of the world by running calculations en masse, offering thousands of simulated examples from which to explore and learn. But computers cannot be relied upon to build models of the world or to make sense of their answers. Those bookends of the process are more appropriately handled by humans. Our skill resides in the ability to evaluate the assumptions behind each model, the reliability of the specific inputs that are fed into them and the plausibility of their outputs. That requires a conscious engagement with what goes into, and what comes out of, those calculations.

What goes in: models

Machines will calculate at the behest of humans, but they have no way of grounding those sums in meaningful contexts, which can result in absurd outcomes. Only an automated alert system could assign a negative age to pregnant mothers,[17] and only an algorithmic pricing model could lead to two booksellers charging millions for an obscure textbook about flies.[18] In each case, the computers faithfully executed the commands issued to them but were unable to rein themselves in when their answers spiralled out of control.

Machine learning programs promise a little more, as they 'learn' their models from data. They produce more than just numerical outputs: they can (among other things) label images, respond to speech cues, play board games, drive cars and engage in text chat. But these programs, too, can err in the most unfathomable ways. The list of examples of 'AI gone wrong' is growing all the time, with anecdotes ranging from the amusing to the abhorrent. Amazon's voice assistant, for instance, decided it was appropriate to order a $170 dollhouse and four pounds of cookies simply because six-year-old Brooke Neitzel asked for them,[19] much to the chagrin of her parents. You can find leading-edge image recognition software that mistook the Star Trek logo for a sea slug.[20] And at the more worrying end of the spectrum is the Microsoft chatbot that spewed racist bile after being trolled on Twitter, and had to be pulled within a day of going live.[21]

We'll take a closer look at the workings of machine learning in the coming chapters. Suffice it to say, a major limitation is that these programs have no concept of the world against which to benchmark their calculations. They have no notion of what may count as amusing or abhorrent. It is for humans – disgruntled parents, diehard Trekkies and Microsoft developers – to put the reins on senseless machine behaviour. These programs, after all, are our own creations. Every model, however simple or complex, is shaped by the choices of its designer. It is the human, not an algorithm, who is really doing the modelling, by choosing functions and parameters* that they think will give reasonable approximations of the world. A computer will only think within the scope of the choices specified to them. The blame for inexplicable behaviour must lie with us, and by the same token, we must assume responsibility for sense-checking the consequences of every choice we feed into our models. That means not being overawed by the processing might of computers. It means not ascribing infallible traits to them. And it means trusting in our common sense to keep automated models in check.

What goes in: inputs

A model lives and dies by the data that it feeds off. The 'garbage in, garbage out' principle that data scientists abide by has held true as long as humans have sought to estimate difficult-to-grasp quantities, as a classic pair of examples illustrates.

In 250 BCE, Eratosthenes, chief librarian at the Library of Alexandria and the 'father of geography', wanted to calculate the size of the Earth.[22] Lacking the tools to carry out precise measurement, Eratosthenes ingeniously devised an estimate. He knew that the city of Syene was roughly 5,000 *stades* – the equivalent of around 925 km – south of Alexandria, and that it lay on the Tropic of Cancer. That meant that on noon of

---

*The term 'hyperparameter' may be more appropriate here. In machine learning contexts, it refers to the aspects of a model, say, a neural network, that need to be specified before the machine can undertake learning.

the summer solstice, the sun was directly overhead. A vertical pole in Syene would therefore make no shadow at that point in time. Meanwhile, in Alexandria, Eratosthenes placed another vertical pole at noon on the same day and measured the angle it made with the ground from the shadow it cast. This angle was seven and a half degrees, approximately a fiftieth of a circle. Using basic geometry, Eratosthenes could now place Alexandria and Syene on the Earth's circumference, separated by a distance one-fiftieth of the whole way around. Since that distance is 925 km, his estimate of the Earth's circumference (the 'size of the earth') came to 50 × 925 km, or 46,250 km. Modern calculations put the true value at 40,075 km, which means Eratosthenes was accurate to within 15 per cent – a valiant effort when you consider that his map of the world, mostly derived from tales of travellers, accounted for just 8 per cent of the world as we know it today.

Seventeen centuries later, Florentine mathematician and geographer Paolo dal Pozzo Toscanelli pitched a proposal to the Portuguese court to sail west to the fabled Spice Islands. Toscanelli had developed his proposal after consulting with Niccolo Conti, the first Italian merchant to return from the Far East following Marco Polo's journey. Toscanelli never made it to sea himself, but his proposal inspired Christopher Columbus's voyage in 1492.[23] Unfortunately, Toscanelli's map had the Earth's circumference as around 30,000 km – a gross underestimate. When Columbus made landfall in America, he thought he had reached Japan, not realising he had actually stumbled upon an unknown continent lying between Europe and Asia. Toscanelli's estimate broke down not because his calculations were at fault but because he got one of his inputs (the Earth's circumference) horribly wrong. Garbage in, garbage out.

It is difficult to secure a reasonable estimate from spurious inputs (you would have to consider yourself lucky if you did). A major area of concern for machine learning practitioners is the prevalence of errors within the datasets they train their models on. A study of ten widely used datasets for computer vision, published in March 2021, found an error rate of 3.4 per cent.[24] This is hardly encouraging when you consider that these datasets form the basis of so many of today's leading image recognition tools. A 'data-centric' school of thought is emerging within the AI space that emphasises the importance of having good data, and the danger of placing faith in sophisticated models without it.[25] In fact, the performance of some of the most complex models is found to be diminished once those errors are corrected for. Bad data has blinded our sense of what makes a good model.

What comes out: weighing up estimates by thinking in ratios

When dealing with large numerical outputs, we can use our knowledge of the world to create benchmarks of what constitutes a reasonable guess. One way to describe 'largeness' is in terms of *orders of magnitude*, which are just different categories of size. The orders are usually defined in terms of powers of 10: 1, 10, 100, 1,000, and so on. The idea is that each category represents the same-sized leap above the previous one (in this case, a factor of 10) and a good estimate is one that falls in the correct category. Even if our estimate is wrong, by getting its order of magnitude right we limit the scale of our error.

Suppose your lunch bill hovers around £10, compared with £100 for a grocery shop, £1,000 for monthly rent, and £10,000 for the car that has always felt like an indulgence too far. Those values are not precise, but they serve as benchmarks against which we can classify everyday expenses. When planning options for your next holiday, we can consider the cost in terms of a grocery shop (as in the case of a one-night getaway) or a month's rent (as in the case of time abroad). The standalone act of selecting the right order of magnitude ensures that our errors are contained within reasonable boundaries, which is often fit for our needs.* This is especially true when working with large numbers (and it explains why astrophysicists are the subject of so many memes that caricature their desire just to get their figures in the right ballpark).[26]

Orders of magnitude rely on comparison by *ratios* rather than absolute differences – in the cases above, each category is separated by a factor of 10. It turns out that we naturally perceive numbers in terms of ratio; our approximate number sense is based on them. Our estimation abilities become fine-tuned with age. With enough experience, we can glean that a set of eight objects is larger than a set of seven through observation alone, without recourse to any counting mechanisms. At all ages, however, our ability to estimate declines for larger numbers: we find it harder to distinguish a 22 kg weight from a 21 kg weight than a 2 kg weight from a 1 kg weight. This is known as the *Weber–Fechner effect* and it will be familiar to you if you have ever perceived time as passing by more quickly the older you get: the passage, say, from thirty years

---

*If we require even more accuracy, we can create boundary values that tie our calculations to an even narrower range of plausible values. For instance, I know that my monthly phone bill falls into tens of pounds – this order of magnitude estimate tells me it will never fall below £10 or climb above £100. I know that the actual cost varies depending on my usage, and if I want a more precise range of estimates, then I can take the lower and upper bounds as £30 and £50 because I know that my bill almost never falls outside the £30–£50 range.

to thirty-five seems to be much more rapid than the passage from ten to fifteen. To a thirty-five-year-old, five years is a mere seventh of their lifespan – over in a flash. To a fifteen-year-old, the same amount seems like an age because it accounts for a third of their total lifespan (actually higher still since the first few years of life escape our memory). Various biological explanations have been put forward as to why this effect takes hold – one theory suggests that it relates to our slowing metabolism as we age (younger people have faster heartbeats, giving the sense that things are happening more slowly around them).[27]

To see the effect in action, very quickly decide where you would place 1,000 on the following number line:[28]

●————————————————————————●
1   1,000,000

You have almost certainly figured that 1,000 must be closer to 1 than to 1,000,000 – but how close, exactly? The chances are that you placed 1,000 some noticeable way along the line. Does it surprise you that the correct placement is somewhere on the dot just above the 1? This makes sense when we pause to consider that 1,000,000 is made up of a thousand copies of 1,000; as a result, we should only travel one thousandth of the way along the line. But because we intuit the relative size of numbers in terms of their *ratio* rather than their *difference*, we tend to think of 1,000 as much closer to 1,000,000 than it actually is (you could say 1,000 is the *same ratio apart* from 1 as it is from 1,000,000). The fact that whole numbers are evenly spaced only becomes ingrained in us through formal study: somewhere between the ages of six and ten, we realise that the difference between 8 and 9 is the same as the difference between 2 and 3. And even then, as you've just experienced, we naturally retreat to ratios.

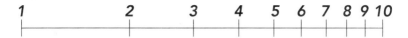

We perceive the number line on a logarithmic scale, with
numbers more squished together than they actually are.

A more precise way of describing this effect is to say that
we have a *logarithmic number sense* – that is, we perceive the
number line on a logarithmic scale. This is the same logarithm
which John Napier introduced to the world to make calculations easier, and which led to the slide rule, where the distance
between two numbers represents the ratio between them, and
the distance between successive values therefore shrinks as
the numbers get larger. Napier's mathematical innovation, it
appears, marries up with our deep intuitions about number.

Our affinity towards ratios over differences has consequences, good and bad. There are times where it makes good
sense to think in ratios. On other occasions, doing so can
distort our perception of the world.

Suppose you are considering launching a new line of cars
in the UK. You need to determine how many cars, roughly, are
sold there each year. One of your researchers returns with an
estimate of 100, another with 5 million. The correct answer,
you learn, is around 2.5 million. Which estimate was closer
to the mark? In absolute terms, both estimates are about as
accurate as one another: the first is out by 2,499,900 compared
with 2,500,000 for the latter. Yet common sense tells you that
the first estimate is nonsensical. In terms of ratios, the first estimate is actually 25,000 times below the actual value, whereas
the second estimate is just two times above it – still not as
accurate as you might hope, but at least within the realms of
reasonable.

Now imagine your new line has been a roaring success and

you are handed a pay rise of £10,000. It sounds like good news, but that isolated figure lacks context. To evaluate how good the news really is, we need a baseline figure: your current salary. A £10,000 rise on £30,000 represents a 33 per cent increase – well worth celebrating. The same increase on a weighty £500,000 salary represents just a 2 per cent rise – probably not enough cause for celebration; you may even take offence at the paltry recognition of your efforts. Our judgement is informed by *proportional* increase rather than increase in absolute terms – ratios rather than differences. It's for the same reason that you probably don't tip all waiting staff the same amount. Restaurants commonly express the service charge as a percentage of the bill, which is an implicit acknowledgement that the market value of their staff is in proportion to the expense of your meal. Ratios trump differences.

Marketers often take full advantage of our predilection for ratios: tagging £5 onto a £100 shopping bill always seems easier to justify to ourselves than lumping the same item onto a £10 shop. In terms of ratios, it is the difference between 5 per cent and 50 per cent of the original amount. We would do well in such scenarios to resort to absolute differences; we will be out of pocket just the same.

Overcoming exponential bias

Our 'logarithmic' sense of number has much to answer for, including our propensity to downplay pandemics in their early stages. During the initial phases of Covid-19, the term 'exponential growth' was on the lips of epidemiologists, journalists, even politicians. It was established early on that cases were doubling every three or four days. 'Exponential growth' is appropriate here; it relates to any situation where an amount *multiplies* by the same amount at regular intervals. It means that growth is itself accelerating. Exponential growth is a whole level up from steady *linear* growth, where the same amount is merely added at regular intervals – the volume of water you have consumed, for example, grows linearly (assuming you drink about the same amount each day).

As Covid-19 cases remained in the tens and even hundreds, the sense of alarm was muted in many regions. Many forecasts presented a gross underestimate, suggesting cases would plateau in the low thousands (which, within a population of millions, would represent relatively low risk). Those intuitions were defeated as cases exploded into the millions and deaths into the tens of thousands (and beyond). The public – and, in too many cases, our politicians – were seemingly caught unaware through multiple waves of the pandemic. What went wrong?

Our trouble in grasping exponential growth is not new. According to a classic Indian legend, when the brahmin Sissa ibn Dahir was offered a prize by the tyrant King Shihram for inventing an early form of chess, he humbly asked for a single grain of wheat to be placed on the first square of the board, then two grains on the second square, four grains on the third and so on, doubling each time up to the sixty-fourth square. The king was glad to oblige, amused even by the brahmin's simple request. That was until he realised at around the thirty-second square that the amount of grain needed for the next would exceed all the food in the land (the exact number after sixty-four squares is 18,446,744,073,709,551,615). If more morbid renditions of the tale are to be believed, the king proceeded to behead the brahmin for such inconvenience.

The king suffered from what psychologists term 'exponential growth bias', and it's a bias that afflicts most of us,[29] including people with higher levels of education. Another example: suppose you stepped out for your daily walk and were asked to take thirty steps. You can reliably estimate where you would end up – the newsagent perhaps, or the house of the neighbour you're keen to avoid. But now suppose you decided to double the length of each stride (indulge me). The first five steps, for instance, would get you as far as 1 + 2 + 4 + 8 + 16 = 31 normal-sized steps (in sight of that annoying neighbour). Where would you end up after thirty super-sized steps? The end of your city, perhaps? The country? In fact, assuming each normal-sized step is one metre, you'd travel the whole circumference of the Earth twenty-six times over. It seems implausible, but the arithmetic cannot be disputed – the nature of exponential trends is such that numbers eventually defy our gut instincts.

Exponential bias can have dire consequences. It can hurt our bank balance by inhibiting our understanding of compound interest. If you invest £1,000 at an annual interest rate of 5 per cent, how much will you have earned in forty years? Most people do not realise that the savings amount to an excess of £7,000 and, by opting out of such schemes, they lose out on a retirement buffer.[30] The consequences can even be deadly, and not just for humble chess enthusiasts. Research shows that the extent of this bias predicts the seriousness with which we take pandemics. A stronger bias makes us less likely to take precautionary measures such as social distancing and mask-wearing.[31]

But why does exponential bias persist even in those early stages of a phenomenon, when we can see it unfolding before our very eyes? An evolutionary perspective is that, until very recently, civilisation has progressed at a steady clip. For millions of years, our lives have been slow, steady and predictable – linear, you might say. So when we are confronted with exponential trends that do not match up to our lived experiences, we cannot help but interpret them as linear.

Our logarithmic number sense is also at play. Even as case numbers multiply at regular intervals, we perceive each successive 'leap' as the same size. The leap from thirty-two cases to sixty-four feels the same as the leap from sixteen to thirty-two (just like the slide rule) – our perception is one of consistent growth. However, the growth is consistent in terms of the *ratios*, not in terms of the actual differences. In effect, our logarithmic sense knocks our perception of growth one scale down: we perceive exponential growth as linear.

Another way of thinking about this is that we tend to exaggerate the size of small numbers; we feel that thirty-two is much closer to sixty-four than it actually is. So when we imagine future case numbers and we think about the results of a succession of doubling, we end up with a gross underestimate. Remember where we placed 1,000 on that line from 1 to 1,000,000? We perceived this modest amount to be much larger than it actually is. When left to intuition, even our 'worst-case' projections are way below the mark because we already perceive 'low thousands' to be staggeringly high.

We can visualise exponential growth bias with two charts that each show the early rise in Covid-19 deaths in the United States. Both charts represent the same data – the difference is in the scale of the vertical axis. In the first chart, the vertical scale is linear – numbers are spaced apart equally. The growth is clearly exponential; we can literally see that the rate at which deaths are increasing is itself increasing. Now look at the second chart: the vertical scale is logarithmic – the distance between numbers represents their ratio, not their difference. The graph now resembles a straight line (near enough). This representation is a popular choice because it makes it possible to plot

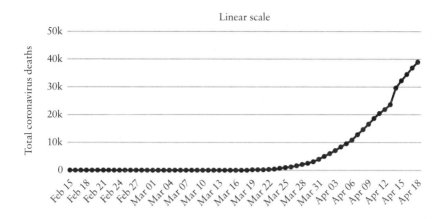

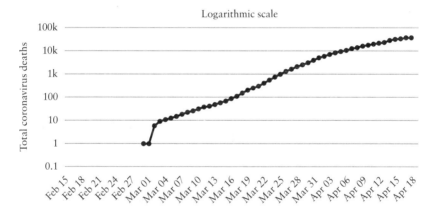

Covid-19 deaths in the United States between 15 February 2020 and 18 April 2020, plotted on a linear scale and a logarithmic scale. The logarithmic scale gives the appearance of linear growth.

large values (especially in situations where growth stretches into millions and beyond).

Both charts are equally valid, and both were given airtime in government briefings and the mainstream press. The second one, by invoking a logarithmic scale, gives the impression of steady growth. One study suggests that people who were exposed to the logarithmic scale version were more likely to misinterpret the data and to underestimate their projections of future case numbers.[32] The straight-line appearance of that graph fuels the misperception that cases are rising at a constant rate (when what that line actually signifies is a steady *doubling* of cases every few days). It looks more reassuring than its sharply rising counterpart. When scores of politicians and pundits played down the threat of the pandemic early on (even Elon Musk, on 19 March 2020, projected 'close to zero new cases by the end of April' in the United States, in a now infamous tweet), they may well have had visions of linear growth in mind.

Covid-19 has given us pause to think about how we perceive numbers, especially larger ones. Getting a handle on exponential growth has become a survival skill even outside of pandemics, as our lives are increasingly governed by exponential trends. Technology is growing at an exponential rate (remember Moore's Law – processing power is doubling roughly every eighteen months), resulting in an explosion of computing power and information. If we fall victim to our logarithmic sense, then we are sure to underestimate the long-term impact of technology. Consider how advanced today's technologies are compared with two decades ago, when the internet and social media were in their infancy and smartphones had yet to go mainstream. The nature of exponential growth means that the technologies of the 2040s will be even further apart from today's state of the art. Just as pandemics have an unnerving way of spiralling out of control, technology itself is bound to evolve in ways we cannot anticipate. This is what the futurist Roy Amara alluded to when he warned that 'we tend to overestimate the effect of a technology in the short run and underestimate the effect in the long run.'[33]

The way to mitigate our exponential growth bias is to think of our future trajectory as an amplified version of past trends. One proven way to get a better handle on pandemics is to focus on the time it will take for cases to reach large thresholds, rather than just looking at daily figures.[34] Visualising trends, and paying attention to the scales we use, can also help to illuminate these projections of the future.

The pandemic is a reminder that our perceptions of number are innately imperfect, but they can combine with the outputs of computers to help us make better sense of the world.

Escaping the Chinese room

Precise as calculation may be, as a standalone activity void of context it cannot be considered intelligent. In his 'Chinese room' thought experiment,[35] the philosopher John Searle called out mindless information processing by asking you to imagine that you are in a closed room. A passer-by slips a sheet under the door containing a list of questions written in Chinese. The bad news is that you do not understand a word of Chinese (you are the only one who knows this). The good news: in the room, there is an instructional manual written in English that shows you, step by step, how to convert the given characters into a set of new characters that correspond to the answers. The passer-by receives your answers and is convinced you understand Chinese because you are capable of answering questions in the same language in which they were posed. Unbeknownst to her, despite your proficiency in this task, at no moment did you possess even an iota of understanding of Chinese. You are unable to assign meaning or context to what you can only perceive as squiggles. The appearance of intelligence does not render your actions intelligent.

Searle was taking aim at computers; he was contesting the very premise of AI by demonstrating that actions that appear intelligent on the surface can unfold without any conscious effort or comprehension of the task. Searle could just as well have levelled his critique at humans, who default all too easily to performing calculations without consideration of what the numbers represent. We are not invulnerable to performing senseless calculations, a fact amusingly illustrated by Kurt Reusser in a 1988 study where he posed the following question to a group of schoolchildren:[36]

There are 125 sheep and 5 dogs in a flock. How old is the shepherd?

There is, of course, no way to discern the age of the shepherd from the size of his flock. No subtle trick or manipulation can rescue this problem from its absurd framing. Yet Reusser showed that three-quarters of the schoolchildren responded to the shepherd problem with a numerical answer. The students cannot be criticised for their display of mental arithmetic; the calculations are faultless. But in pursuit of a definitive answer, students neglect to evaluate the *relevance* of their calculation.

This chapter should give us encouragement that humans are able to escape the Chinese room: we've developed effective safeguards to protect against mindless calculation. After all, despite our natural difficulty in intuiting exponential growth, many people can reliably model pandemics, foresee the rise of new technologies and cash in on investment schemes. Our knowledge of the world is what helps keep our calculations grounded; it is how we overcome our numerical blind spots.

Humans are not unique in possessing an innate sense of number. The ability to quickly identify bountiful food sources, or to assess how many predators are looming close by, confers evolutionary advantages. Quantitative skills can be found across the animal kingdom, at least where small amounts are concerned. Honey bees can track the number of landmarks as they search for food, female lions are more likely to avoid fights with nearby intruders when they hear three roars instead of one, spiders spend more time hunting on their webs when there are more prey to be found, and crows can conceive of zero as a quantity that is close to the number 1.[37] Rats can distinguish 2 + 2 from 3, while chimpanzees have some intuitive grasp of fractions, and Karen Wynn's findings for infants have also been shown to hold for dogs and rhesus monkeys.[38] What sets humans apart is our capacity for language and abstraction, which allows us to formalise our concepts of number, handle larger quantities and develop theories along the way. Animals may have a sense of 2 and 3, but only a human could tell you that 2 + 3 = 5, or that all three of these numbers are prime.

Our command of numbers, and of mathematics, rests on our ability to represent knowledge – whether knowledge of the world or knowledge of a more abstract kind – in powerful ways. Knowledge representation remains a longstanding challenge of AI, and it is the focus of the next chapter.

2

REPRESENTATIONS

*The dogness of dogs, how mathematicians paint
ideas, and the blind spots of computers*

Mathematicians are misunderstood people. Despite under-
taking some of the most creative mental feats, it is mistakenly
assumed that they while away their days pushing symbols
around.

An overreliance on symbols traps us in the narrowest, dullest
view of what intelligence is. Humans possess an arsenal of
mental tools for making sense of our most complex thoughts.
When literally viewed the right way, mathematics is the perfect
case study in how, unlike machines, we're able to represent
knowledge in the most diverse and vivid ways.

How machines see the world

If you have any ambition to develop AI, you have to reckon
with how your intelligent being will see and make sense of
the world. This has proved to be one of the field's enduring
challenges. Early efforts to develop AI were based on a *sym-
bolist* paradigm, which states that we can encode the world's
objects as symbols and model their behaviours with logical
rules for manipulating those symbols. In this view, intelligence boils down to long lists of hard-coded instructions. To
replicate intelligent human behaviour, the thinking goes, you

simply specify all the rules that expert humans employ when making decisions. This approach was the basis for IBM's chess playing program, Deep Blue. It turns out that you (or rather, a machine) can best a grandmaster by compiling a database of hand-crafted chess moves, taken from expert human players, and then sift through its myriad options, selecting the one that ranks highest according to some scoring function.

But for many problems, symbol crunching only goes so far. This was first seen in mathematics, which has repeatedly found itself in the crosshairs of AI. The rule-based General Problem Solver (GPS), developed in 1957 by Herbert Simon and Allan Newell, is widely considered the first AI program.[1] The GPS had built-in knowledge of various problems, as well as general strategies for solving them. It could solve a range of maths problems that could be precisely stated in terms of symbols (it could also solve certain types of word puzzles and play chess). The GPS proved not to be all that general, however, when it came up short on problems that evade strict symbolic definition.

AI researchers encountered the same challenge in other domains. Say you want to create a machine doctor. No problem, you might think – just arm it with meticulously encoded rules that have been specified by expert human diagnosticians. When a patient presents with symptoms and medical test records, apply those rules and await an automated diagnosis. Alas, any doctor will tell you that no set of rules, however large, can account for the sheer variety of ailments they are presented with. Doctors rely as much on intuition, honed through years of experience and accumulated wisdom, as they do on hard-and-fast rules for diagnosis.

The painful lesson for early AI practitioners was that the world is far too large and complex to grasp with fully specified rules.[2] Much of our knowledge of how the world works is tacit and cannot be formally stated. Think of how you learned to ride a bike – you almost certainly did not rely on an instruction manual as much as your sense of balance. Some of our most fundamental, in-built human traits, like emotion, intuition and common sense, are also the hardest to put down in words. As the philosopher Michael Polanyi put it, 'we can know more than we can tell.'[3]

Intelligence also rests on the ability to *learn* through experience and interaction with one's environment. To ride a bike, we hop on multiple times, we stumble and we learn from every error, honing those senses until finally we have the confidence to retire our stabilisers. This is the insight at the heart of *machine learning*, the dominant approach of modern-day AI applications. In a bid to unshackle computers from a rigid knowledge base, and to capture some of those subtler elements of thinking, the idea now is to allow computers to 'learn' from data inputs. A model is set up to represent the situation at hand, data are fed into that model, and an algorithm crunches that data to determine the precise shape of the model (its 'parameters'). The machines are 'learning' in the sense that their parameters evolve as they receive more data – their models become more accurate (that's the hope, anyway).

Many of machine learning's most promising approaches have taken inspiration from the human brain. The subfield of *deep learning*, for instance, deploys models that are loosely based on the brain's neural networks.[4] In the human brain, neurons fire up when they receive strong enough electrical signals from other activated neurons. The algorithms behind 'artificial' neural networks try something analogous by adjusting the weights between neurons as data comes in. Another subfield, *reinforcement learning*, draws on behavioural science by using reward-and-punishment schemes in order to incentivise machines to make effective choices.

The theory underpinning these approaches has been around since the 1960s, but it was only decades later, with advances in processing power and the availability of large datasets, that their potential was activated in areas such as image recognition and autonomous vehicles.

The most high-profile example to date comes from the lineage of Google DeepMind's Go-playing machines, starting with AlphaGo. A rule-based approach was never viable for Go because no single database could account for its astronomical range of possible scenarios. Instead, the DeepMind team have deployed deep learning and reinforcement learning in tandem to produce stunning results.[5] Unlike Deep Blue, AlphaGo (and its successors) get better with experience. Whereas Deep Blue used the exact same logic in every game, AlphaGo learns from every move and every game, self-correcting as it goes. It updates the parameters of its models, discovering fine-grained patterns in gameplay as it acquires deeper expertise.

On the surface, it appears as if machines are inching towards human ways of thinking. In fact, humans can no longer compete with machines in Go and, what's more, the machines play with a style and grace that has enthralled aficionados. Machine intelligence seems to look more human-like, even superhuman-like, than ever.

The dogness of a dog

Upon closer inspection, however, machine learning programs behave nothing like humans. A machine treats all objects as vectors: as strings of numbers. When a machine learning algorithm is fed training examples – images, text, the positions on a Go board – it represents each of them as a vector and performs mathematical operations to find a function that best describes them (a procedure known as *optimisation*). A simple example that you may recall from school is the 'line of best fit' where, given a set of points in the plane, we draw the line that passes through them most closely. The algorithms of machine learning boil down to figuring out what that line should be. They are rather more complicated – for one thing, they usually involve millions, even billions, of parameters that, to humans, are impossible to visualise (but that we have the mathematical tools to deal with). Many of these parameters are derived from others using complex calculations – layers upon layers of abstraction that make them difficult to interpret.

Driving these approaches is a very particular *mathematised* view of the world, steeped in a handful of techniques from fields like statistics, linear algebra and calculus. It's a worldview that is decidedly narrower than that of humans, even of mathematicians for that matter.

Consider a simple everyday example: how do you recognise a dog when you encounter one? As a child, you would have roamed your environment, absorbing all kinds of visual and audio cues, until those curious four-legged animals piqued your interest. Your parents might have pointed them out by saying: 'Look, there's a dog.' They may have repeated this only a few times, but soon enough you would have assimilated *dogs* into your mental models of the world. You learned to associate them with parks, where they popped up in large numbers, as well as with bones, balls and other objects that dogs interact with daily. Through observation and physical interaction, you would also have discovered that dogs could be patted, fed, chased around and washed. Your knowledge of dogs doesn't reside in a single place; you possess thousands of complementary models, fed by

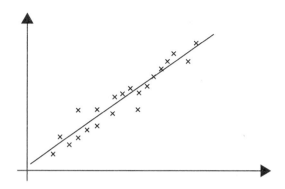

A line of best fit – a basic illustration of what machine
learning algorithms are designed to achieve.

your senses, that arrive at a consensus of what *dogness* entails.[6]

The human brain learns continuously by situating objects
within the context of our prior experiences. Our worldview
updates one incremental step at a time as we relate new experiences and problems to old ones. A machine learning program,
in contrast, is hungry for all the data it can get and processes
each input indiscriminately. It needs to be shown thousands of
images, some labelled as dogs and others not, before it develops
a reliable way of spotting one.

Even then, a computer does not 'see' a dog in the same literal
sense that humans do. The program is essentially performing
a calculation that compares one constellation of pixels with
another (each pixel is represented by a number that denotes its
brightness). If one photograph's pixels are, in numerical terms,
'similar' to other images with a dog, the system will 'intelli-
gently' guess that the new photograph is of a dog.

An artificial neural network goes further. It is designed as
a hierarchy of neuron layers, each one seeking out higher-level
detail that combine to form an overall image of a dog. But where
humans have a vivid concept of dogs because our brains are

attuned to picking out defining features, machines are drawn to microscopic detail. Mathematician Hannah Fry describes the contrast: 'It's not looking for a measure of "chihuahua-ness" or "Great Dane-ishness" – it's all a lot more abstract than that: picking up on patterns of edges and light and darkness in the photos that don't make a lot of sense to a human observer.'[7]

Once again, the results are impressive: a state-of-the art algorithm can be trained to identify breeds of dogs more reliably than humans can. But the computer, seeing everything as vectors, has no vivid concept of what a dog actually is, or its relationship with other objects. For all the machine knows, the dog could be something you drink or get married to. Similarly, the Go-playing machines of DeepMind have no conception of Go beyond the abstract manipulation of numbers.

What machines do not yet know

Computers are adept at processing data, and finding patterns among them, but they do not 'learn' with context or meaning. Machines have no conscious awareness, no temporal sense of the world (that is, they cannot link different sequences of events together), nor any models for causal reasoning. And whereas the thinking processes of those earlier expert systems were plain enough (because they imitated humans), the 'black boxes' of high-performing machine learning algorithms are difficult to probe; this can catch us unaware when they behave in unexpected ways.

The arcane methods driving machine learning make for some dubious behaviours. One state-of-the-art program that accurately distinguished wolves from huskies turned out to be basing its classification purely on whether snow was present in the images.[8] Another program that accurately picked up melanoma simply went on surgical markings.[9] And a model for predicting people's ages from their pictures was strongly influenced by whether a given person is smiling or wearing glasses (if only ageing really worked this way).[10]

Because they possess such a narrow worldview, machine learning programs are often found to be brittle when faced with a situation that deviates from the specific data they have 'learned' from, such as when imperceptible noise is added to an image or sound. In fact, machine learning can go awry even when such distortions are blatant. One neural network was able to reliably recognise bananas, but when the researchers added a small sticker of a psychedelic toaster next to the banana, the algorithm classified the image as a toaster.[11] Only a human, it seems, could tell you that the image is still of a banana, with a funny-looking sticker by its side.

These problems aren't limited to vision. In natural language processing, when OpenAI released its language-generating system GTP-3 in 2020, social media was abuzz with excitement at its apparent ability to produce a wide range of texts. The premise of these systems is that you can enter some text – say, a few paragraphs – and it will continue the passage in the same style. In this case, it is the letters and words of natural language that have been vectorised.

It doesn't take much to expose GPT-3's lack of comprehension. Cognitive psychologist Gary Marcus (known for his rebuttals of AI-induced hype) presented a list of prompts that he issued to GPT-3,[12] which each resulted in responses that, to the human reader, are patently absurd. In one example, Marcus started with: 'You poured yourself a glass of cranberry juice, but then you absentmindedly poured about a teaspoon of grape juice into it. It looks okay. You try sniffing it, but you have a bad cold, so you can't smell anything. You are very thirsty.' To which the system replied: 'So you drink it. You are now dead.' To the comprehending mind, there are no fatal prospects to sipping the cranberry juice – that really isn't the point of the passage. But GPT-3, which was fed an unfathomable amount of information during its training phase, somehow related the passage to tales of poisoning. The mere act of extending a passage is a step removed from genuine understanding – it requires holistic thinking about how multiple ideas fit together, as well as knowledge of linguistic structures such as syntax, phonology and semantics.

We're increasingly relying on machines to look at the world, to read and listen to information, to summarise what they know, and to make snap decisions on our behalf. In our rush to deploy these technologies, we risk ascribing to them perceptual qualities that amount to little more than derivative number crunching. As they become more pervasive in everyday use, we need ways of making their thinking visible – the data they've been trained on, the optimisation techniques at work – so that their errors can be exposed and rooted out.

Humans are not immune to cognitive errors (quite the understatement, as we'll see in the next chapter), but we can at least express our ideas in ways that elucidate our thinking processes. Doing so has the added benefit of enabling us to hold machines accountable by demanding that their own 'learning' mechanisms are made more transparent.

Hybrid thinking

Neither the rule-based machines of the past nor their data-hungry successors capture by themselves what it means to truly know something. There's a growing recognition that AI systems will have to integrate both rules and data.[13] The human brain is an exemplar of a hybrid thinking system that blends hard-coded knowledge with learning algorithms. We are not the 'blank slates' suggested by philosopher John Locke, waiting to be inked by our environment. In the previous chapter we've already glimpsed deeply innate traits like an approximate number sense, which is one of many pre-configured intuitions we bring to the world. Nor, at the other extreme, can our DNA fully specify all the types of knowledge we are able to create. The cognitive neuroscientist Stanislas Dehaene demonstrates this constraint with a back-of the-envelope calculation, noting that our DNA contains around 6 billion bits – just enough to fill a CD-ROM, and nowhere near sufficient to account for the 100 terabytes he estimates for our brain capacity.[14]

For Dehaene, the human brain is 'the result of a compromise'. As he puts it: 'we inherit, from our long evolutionary history, a great deal of innate circuitry (coding for all the broad intuitive categories into which we subdivide the world: images, sounds, movements, objects, animals, people …) but also, perhaps, to an even greater extent, some highly sophisticated learning algorithm that can refine those early skills according to our experience.'[15] The result of this compromise is a phenomenally diverse repository of models and techniques for representing ideas. As babies and infants, we rapidly form an understanding of the world by interacting with it, developing representations of people, objects, ourselves. Every life experience iterates and expands on our ways of seeing the world. For humans, learning is a kind of combinatorial play. Using a swathe of linguistic tools such as words, metaphors, symbols and pictures, we join together existing pieces of knowledge in novel ways to arrive at new concepts.

The next chapter will look at the causal mechanism of logic that helps us to rigorously tie together concepts in order to derive objective truths. Before that, we look at how mathematicians seize upon different knowledge representations to make sense of complex ideas.

Mathematics has an odd coexistence with AI. Early attempts to automate mathematical problem solving fell short because mathematics doesn't boil down to symbol manipulation. On the flipside, recent efforts to solve intelligence using a small handful of mathematical techniques fail to capture our diverse and subtle ways of seeing the world. Mathematics can be a standard bearer for understanding how to express ideas, but it first requires that we embrace the sheer breadth of representations the subject gives rise to.

Abstraction and language in mathematics: the birth of a number system

Humans are, as far as we know, unique in our ability to ascribe words to signify particular amounts. In some cases, the resultant vocabulary is tied to specific objects that are used in everyday life. For instance, early Aztec languages reserved terms to describe 'one stone' and 'two stones', and languages of the South Pacific do likewise for 'one fruit' and 'two fruits'. It would be painfully exhaustive to have to contrive new words for every object type. It is more economical to describe a quantity without reference to any specific object. This is the first abstraction that takes humans beyond our instinctive attachments to quantity, which in turn helps us acquire a precise grasp of larger amounts.

Consider the number *three*. If you were to scurry around the breadth and depth of the universe, you would never encounter 'three' as a physical object. Instead you would find groups of objects that appear to have a certain *threeness* about them.

In the image below, the quantity of apples appears to be the same as the quantity of ducks. We could pair each with a duck in perfect correspondence, which tells us there is a numerical property common to both groups: they both possess threeness. We use the term 'three', or better still the symbol '3', to denote the property of threeness. It can now be applied to any type of object and, in some very real sense, connects all of those specific instances where three objects arise. Abstraction digs into the most essential characteristics of an object.

In ancient times, we relied on simple counting schemes such as pebbles or fingers to put this abstraction to use. A Neolithic shepherd could measure his flock by matching each sheep against a pebble. After all the sheep are matched, the shepherd might recognise the *fiveness* property of the pebbles, from which he knows his flock also has this fiveness, i.e. he has five sheep. We now have a way to capture, in exact terms, the size of object groups that escape our eyeball estimates. The only remaining bottleneck is language: what scheme of words or symbols might we use to describe increasingly large quantities? And how can we get a handle on large quantities without resorting to counting every individual item in turn? These questions carried everyday importance to early traders, who relied on precisely calculated quantities when bartering goods.

The solution to handling large amounts is to divide them into smaller groups of a fixed size. To compare which of two large piles of grain is larger, for instance, it is now only a matter of counting off the number of groups, rather than the number of actual grains. But what group size should we opt for? The decimal system that is now baked into our everyday use of numbers is predicated on a group size of ten, a choice inextricably tied to our anatomy. Our ten fingers are the most portable of all counting devices, rendering ten the most natural and reliable of group sizes. When we write the number 84, for example, we are denoting eight groups of ten, with four left over. We know immediately that this quantity is greater than 67 because the latter is only made up of six complete groups. Comparing the number of groups (i.e. *eight* tens vs *six* tens) is far less cumbersome than counting every item in turn.*

The grouping approach has the major advantage of scale. When we reach ten groups of ten, we can label this 'group of groups' with a new term, *hundred*. When we reach ten of those, we introduce *thousands*. Each time the quantities threaten to escape our grasp, we rein them in with a new label that keeps them within our walls of perception.

The human choice of *base ten*, which was probably first made by the Egyptians, was a biological happenstance rather than an inevitability. It is a choice that has been challenged repeatedly throughout history. In the eighteenth century, Reverend Hugh Jones championed the base eight *octal* system as an alternative, on the grounds that it appealed to quantities used in the kitchen (40 fluid ounces make a quart, 16 ounces make a pound – both multiples of eight). In this system, group sizes would correspond to eight, sixty-four (eight groups of eight) and so on. The '*dozenal*' base twelve system, which already makes its mark in timekeeping, gained support in the twentieth century and has legions of fans who attest to its power over

---

*Notice how, even when developing the most formal concepts of number, we sought to minimise the burden of computation.

the decimal system. They cite the fact that 12 can be divided by several smaller numbers (namely 1, 2, 3, 4, 6, 12) compared with the number 10 (just 1, 5, 10), which makes arithmetic a lot easier in base twelve.[16] The most ardent campaigner, however, would probably concede that, despite these advantages, the base ten system is far too entrenched to justify any shift.*

Even as the decimal system reigns supreme, we continue to rely on the remnants of other base systems that arose from the choices of civilisations past. The Babylonians were fond of the sexagesimal *base sixty* system, which may also have derived from our finger constructs (each finger is divided into three segments, meaning we can count up to sixty by taking all combinations of finger segments on one hand and digits on the other). Traces of base sixty are all around us. We divide each day into twenty-four chunks known, of course, as *hours*. The Middle Ages saw the hour divided into the *pars minuta prima*, or 'first small part', which denoted one sixtieth of an hour and became known as a *minute*. We divided the minute once more into a 'second small part' comprising a *second*, which seemed a reasonable length to discretely mark the passing of time, as it roughly corresponds to the duration of a single heartbeat or breath. Our biological form factors are never far away from representations of number.† This can also be seen in the adopted

---

*Indeed, there have been campaigns to unify our ways of counting around the decimal system. In 1793, a decree was issued in France that established the *decimal hour*, which would divide the day into ten equal parts rather than twelve. The *decimal minute* and *decimal second* followed naturally. The reign of decimal time was short-lived, lasting around six months.

†The same may be said of computers. The binary system that computers operate within, comprising solely zeros and ones, is predicated *base two*, an engineering choice based on circuit design (early machines were, in fact, decimal).

conventions of the tribes we met in the previous chapter. The Oksapmin people of New Guinea, for example, have developed a *base twenty-seven* number system by drawing on the same number of distinct body parts, starting from the thumb of one hand, up through the nose, and finally onto the little finger of the opposing hand.

The 'unreasonable effectiveness' of mathematics is partly explained by the fact that its own origins lie in mental structures that reflect the most salient features of our environment and of our own bodies. Different bodily and environmental features could give rise to different conceptions of mathematics. My niece is a polydactyl: she shares Anne Boleyn's rumoured anatomical quirk of possessing an extra digit. We delight in the matter with our 'high six' greeting, unique to her among our family members. If all of humanity shared this feature, we might have conceived a number system premised on a group size of eleven. Since it would have originated from the asymmetry of a six-fingered hand, this might have led to fuzzier notions of *a half*. We can teleport this thought experiment across the solar system by asking what concept of number might have been developed by the inhabitants of Jupiter: through occupying a gaseous world of continuous motion, they might even have settled on a more fluid concept of quantity, more in line with our approximate sense of number.

We can only speculate what base system distant aliens might employ, and how their bodily features might inform their own representations of number (if indeed they use numbers at all).[17] If the little green people did ever drop us a line, we would need a shared language convention that cut across our intergalactic divide. In Carl Sagan's sci-fi novel *Contact*, later adapted into a movie, the aliens transmit signals as a sequence of prime numbers: 2, 3, 5, 7 and so on. Prime numbers, remember, are whole numbers greater than 1 that can only be divided by themselves or 1 (that is, they are indivisible into smaller whole parts). Primes remain prime regardless of the base number system used to represent them or where you happen to encounter them in the universe – their indivisibility is *intrinsic*.

Prime numbers give credence to the Platonist notion of mathematical objects as abstract entities independent of human language, thought or practices. But the way we come to understand and appreciate mathematics depends on how we choose to represent it, which, in turn, is rooted in our lived experiences. We relate to mathematical objects by attaching them to our existing worldview and coating them in representations that are familiar to us.

Mental representations and their compressive qualities

Humans are not built for information storage. In the 1980s, Bell Labs researcher Thomas Landauer estimated that the human brain can store around 1 gigabyte's worth of lifetime memories,[18] while biologist Terry Sejnowski's research group estimated the total storage capacity of the brain to be in excess of a petabyte (1 million gigabytes) of information.[19] Such attempts to measure brain storage capacity fall victim to the brain-as-computer metaphor because they presume that neurons are digital and that the brain physically stores information. In actual fact, thoughts and memories are distributed across networks of neurons, as part of a *natural* processing environment. Still, the estimates, however crude, suggest that sheer quantity of information is not the variable humans should be optimising for. Every day, 2.5 quintillion bytes of digital information are created (for comparison, that's over a thousand of those petabytes). This gargantuan number will only rise in accordance with Moore's Law and the proliferation of data-generating technologies such as social media and the Internet of Things. We need ways of compressing information – a skill revered to the extent that some in the AI community believe it is akin to general intelligence.[20]

Compression is a natural human skill. Our visual system patches together snippets of information, constantly filling gaps through approximation and guesswork. The human eye, for instance, contains 130 million photoreceptors, which collectively receive billions of bits of information each second. To manage the load, our visual circuitry reduces billions to millions without much loss of quality (much like photo compression software). Just forty of those (yes, forty) penetrate through to our conscious attention. We see a fraction of what we process, and what's more, our view of the world is bounded by perceptual limits. We only see light of wavelengths between 400 and 700 nanometres, a sliver of the electromagnetic spectrum (which is why, among other things, our eyes are oblivious to microwaves and X-rays). What we perceive as a rich and detailed reality is something of an illusion.[21] The illusion escapes our notice only because our eyes are constantly on the move (three times per second), giving a sense of a complete and integrated picture of the world.

We also perceive the world in broad strokes. When we recognise faces, we do so based on a handful of distinguishing features,* and we are able to identify the same musical piece even

---

*The human brain has a module called the fusiform gyrus that computes certain values such as the ratio between the tip of the nose to the end of the nose and the distance between the eyes. Experiments confirm that we recognise interpret images from just a handful of these values, which makes our recognition skills resilient to subtle changes. It is why we can easily recognise the same face in multiple poses with little fuss.

when it is played in different tones. The mark of a compelling film trailer, book synopsis or business pitch is its overarching premise. When grasping the essence of an idea or object, high-level cues trump detail – and as we saw in the introductory examples of this chapter, it is humans, not computers, that see the forest for the trees.

Our everyday actions rely on this kind of big-picture thinking. The routine task of buying a train ticket or meeting a friend for lunch would soon overwhelm us if we had to contend with every painstaking detail. We navigate the world by arranging our thoughts and actions into hierarchies with several layers of abstraction. We turn minute details into chunky concepts and combine those concepts further still to develop our high-level understanding of things.

To piece together patches of information into meaningful wholes we need strong representations. The cognitive psychologist Anders Ericsson defines mental representations as 'pre-existing patterns of information that are held in long-term memory and that can be used to respond quickly and effectively in certain types of situations'.[22] The operative term here is 'patterns' – information does not exist in isolated pieces.

Psychologists refer to the joining together of information as *chunking*: we chunk information so that there is less of it to grapple with at any one time, which is vital given our limited working memory. If you ever still bother to memorise a phone number, it is likely that you chunk the eleven digits into three groups of 5–3–3: it is easier to manage three separate chunks than eleven tiny bits, and our working memories can only accommodate four to seven objects at once. As we combine sequences of thought together into chunks, we find ourselves capable of holding in our minds arguments of extraordinary complexity. Chunking also accounts for the dazzling feats of elite athletes, who exercise split-second manoeuvres with unswerving accuracy by recognising patterns of play. It explains how musical virtuosos are able to rattle off lengthy pieces by heart. In each case, experts are relying on familiar, repetitive structures.

The mark of experts is not only the number of representations they possess, but also the richness of them. In the mid-twentieth century, Dutch psychologist and chess player Adriaan de Groot ran a series of landmark chess experiments to compare the ways in which players of different ranks assessed the board and planned ahead for their next move.[23] During the experiments, subjects were asked to look at a variety of pre-determined board positions (all plausible gameplay scenarios) and then recall the location of each piece. De Groot found that grandmasters and masters of the game could recall 93 per cent of the pieces, experts recalled 72 per cent and class players just 51 per cent. De Groot's findings were affirmed by later studies conducted by American researchers Herbert Simon and William Chase, who found that for 'real' game positions, the performance of subjects declined in proportion to their chess rating.[24] The higher-ranked players, calling on familiar patterns, were able to rapidly encode chunks when committing positions to memory.* For the chess experts, the knowledge of board posi-

---

*For example, club players easily recognise the fianchettoed bishop on the kingside in a single glance, grasping the six pieces involved as a single collective. Amateurs, lacking these associations, needed to memorise each piece and its location separately. At the highest levels, grandmasters call on a repository containing thousands of such chunks that allow them to identify familiar game positions within three or four seconds. They also recognise the functional relationships between the pieces, rather than the actual positions and spatial relationships. Imagine a chunk of pieces in which a bishop has pinned a knight against the queen; such a position would be remembered as a 'pin' rather as three pieces occupying three distinct board positions.

tions is connected. Chess pieces are not viewed as single units, but as groups that attack and defend. The researchers went further by showing that when pieces are placed randomly, the advantage conferred on higher ranked players disappears. In these scenarios, the board positions no longer carried meaning, and the effort to memorise was just as effortful – one painstaking piece at a time – as for lower-ranked players. The experts had no representations to fall back on, no way of compressing the board into patterned configurations. In one sense, they were reduced to the brute force crunching of machines like Deep Blue – an approach unsuited even to grandmasters.

Experts literally see their craft differently to the way novices do; the distinction lies in the ability to connect bits of information. This is why stories are often termed 'psychologically privileged'; they offer us a natural mechanism for compressing information into manageable chunks. Ed Cooke, a Grand Master of Memory,* suggests that:

> Stories make learning connections easier because they make what happens next feel like it's inevitable. Each item seems to be incomplete without all the others ... the best thing to do is weave the items into a compelling story line. The more that this narrative wraps tightly around the available facts and makes each feel like an intuitive part of the whole, the nearer it will come to pure understanding.[25]

Cooke's ideas on memory are backed by research. Studies show that people tend to read narrative-driven texts twice as

---

*Awarded to people who have demonstrated they can a) memorise 1,000 random digits in an hour, b) memorise the order of ten decks of cards in an hour and c) memorise the order of one deck of cards in under two minutes.

quickly as non-narrative texts, and they also recall twice as much information when tested later on.[26]

Let's return to mathematics. As William Thurston, a winner of the Fields Medal (the highest prize in mathematics), describes:

> Mathematics is amazingly compressible: you may struggle a long time, step by step, to work through the same process or idea from several approaches. But once you really understand it and have the mental perspective to see it as a whole, there is often a tremendous mental compression. You can file it away, recall it quickly and completely when you need it, and use it as just one step in some other mental process. The insight that goes with this compression is one of the real joys of mathematics.[27]

We have already seen the benefits of the decimal number system as a chunking mechanism: grouping objects in batches of ten is convenient because we can count each batch with our fingers. With the group size agreed, the next challenge is to find a way of representing different quantities. For this we use the positional place value system, which originated in India, gained traction with the Arabs and eventually spread to Europe with the help of Fibonacci's 1202 treatise *Liber Abaci (The Book of Calculation)*. When we write the number 137, the *position* of each digit has meaning. Each position corresponds to one of the group sizes, starting with seven *ones* in the rightmost column, then three *tens* and one *hundred*. The compactness of this representation is quite remarkable: with just ten digits, we are able to express quantities of any size. One of those digits, 0, acts as a placeholder to signify the absence of a particular group. For example, the distinction between 1,603 (one thousand, six hundreds, no tens and a three) and 163 (one hundred, six tens and a three) is determined purely by the inclusion of the zero.

The positional number system also gives rise to simple, scalable methods of arithmetic. To add two quantities together, you just need to line up their place value representations and add the group sizes instead: 37 + 22 is a simple matter of combining the *ones* (7 + 2 = 9) and then the *tens* (3 + 2 = 5), giving five *tens* and nine *ones*, or 59. Other operations follow suit, and they scale well to multiple digits. Compared with competing systems such as Roman numerals, which require new symbols every time we strike a new order of magnitude and possess no such rules for adding numbers together, there is no contest: place value is the most succinct of all known arithmetical structures.*

Numbers are especially ripe for compression because they exist within a patchwork of patterns. Consider the multiplication grid: a mainstay of classrooms the world over and the source of many people's negative disposition towards mathematics. At school, many of us were required mercilessly to commit each multiplication fact to memory, up to 12 × 12 (an arbitrary cut-off point and another throwback to the *dozenal* system).[28] When the representation of multiplication is a static collection of symbols, we have to work cumbersomely through 144 disconnected facts. As the French mathematician Henri

---

*As with our choice of decimal, there is nothing deterministic about our place value number system. However intuitive it may appear, its mainstream acceptance has only come about in the past few centuries. The delay owes to a range of factors including consternation around the use of zero as a number that can be operated on, along with fierce cultural attachments to more primitive number systems such as Roman numerals. In the end, the positional number system won because of its unrivalled efficiencies. Yet it remains one choice among many.

Poincaré analogised over a century ago, 'an accumulation of facts is no more a science than a heap of stones is a house.'[29] But what if we change the representation? If we think of every number in our times tables as an area, for instance, we end up with the scaled multiplication grid below:[30]

1	2	3	4	5	6	7	8	9	10
2	4	6	8	10	12	14	16	18	20
3	6	9	12	15	18	21	24	27	30
4	8	12	16	20	24	28	32	36	40
5	10	15	20	25	30	35	40	45	50
6	12	18	24	30	36	42	48	54	60
7	14	21	28	35	42	49	56	63	70
8	16	24	32	40	48	56	64	72	80
9	18	27	36	45	54	63	72	81	90
10	20	30	40	50	60	70	80	90	100

Once one looks past its unusual appearance, the grid will emerge as a source of new insight. It conveys size and proportion instead of just raw numerical outputs. In doing so, it binds together number with geometry and multiplication with area. The mathematician sees shape as well as size and understands multiplication as part of a rich tapestry of ideas. Depending on the representation we elect, multiplication grids can be dull or exciting, creative as well as factual, one-dimensional or teeming with possibility. You might, for instance, consider what a corresponding version of this grid would look like if it included all numbers from 1 to 100. One such version, where each number is visualised as a rectangular array, is shown opposite.[31] Notice how some numbers can't help but shape up as slender tower-like figures because there's no way of dividing them into smaller whole-number parts. We've just stumbled on a new way to visualise the prime numbers.

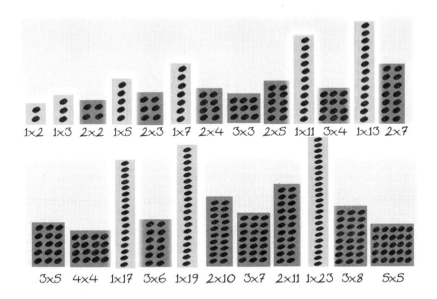

With our new representations, we can intuit that the order in which we multiply two numbers does not affect the outcome: 7 × 9 and 9 × 7 now represent the area of the same rectangle, one rotated onto the other. This single property of multiplication (known as *commutativity*) has a strong compressive effect. In one fell swoop, it reduces 144 separate facts down to 78. We can invoke several other representations that breathe their own life into the multiplication grid. Each representation stitches together those multiplication facts with meaning, allowing us to appreciate numbers as part of an organic structure with several interconnected parts. As educator Paul Lockhart puts it, arithmetic is a form of 'symbol knitting'[32] rather than a set of blunt calculational outputs. Lockhart is referring to numerical symbols, but the same insight cuts across other areas of mathematics, where symbols take the form of letters and represent a wide range of objects. The difference between symbol knitting and what we might term 'symbol pushing' is the difference between the creative mind and the mechanical one.

A picture paints a thousand symbols

When mathematicians speak of beauty, and when they compare the subject to art, they often have symbols in mind. In one study of fifteen mathematicians, researchers used functional magnetic resonance imaging (fMRI) to show that 'the experience of mathematical beauty correlates with activity in the same part of the emotional brain … as the experience of beauty derived from other sources'.[33] The telling aspect of the study is the type of mathematics selected: the mathematicians were shown sixty formulae that they rated as either *beautiful, indifferent* or *ugly*. Implicit in this choice is the fact that mathematicians relate notions of beauty to symbolic representations. The winning formula – that is, the most beautiful of all – was:

$$e^{i\pi} + 1 = 0$$

This configuration of symbols, known as *Euler's formula*, may mean little to the unacquainted. To appreciate its significance, we need to be familiar with each of the five numbers involved and, perhaps most importantly, the exponential function that gives meaning to the term $e^{i\pi}$ – that's the same exponential function that models pandemics and the rise in computing power.[34] The mathematician, having this knowledge at hand, appreciates the formula as the embodiment of economic expression of rich, multifaceted ideas, and is likely, for example, to turn his/her mind to circles and rotations. At their best, formulae exude the qualities of brevity, purity and versatility. A mathematician sees far more than a collection of squiggles, much as Keanu Reeves's character Neo is able to discern people and objects within the green array of symbols in *The Matrix*.

Humans abbreviate at every turn because we tire of repetition. It is no major surprise, therefore, that letters entered the fold as a way of generalising mathematical objects such as numbers. If we possess some belief about the behaviour of numbers, then, rather than futilely attempting to test the infinitely many cases, we can instead represent these objects using a symbol such as x (we could just as well use bananas, but letters are more succinct). The symbol is the most general representation we could hope for and is itself an object that can be operated on. With the use of letters, we can reduce verbosely presented information (those infamous *word problems* so dreaded in school) to symbol manipulation.

Symbols have influenced mathematics from its very beginnings in the form of various numeral schemes, pictograms and the like. Our facility with number is closely tied to our choice of numerals – for example, one reason many Chinese students exhibit a flair for arithmetic is that the Chinese syntax for numbers is highly succinct (the pronunciation of the number 12, for instance, is 'ten two', which does not require new terms or sounds as in the case of the English 'twelve', and which is also the literal interpretation of the place value representation of 12).

Symbols do take some getting used to. The public image of mathematics is a sprawling mess of incomprehensible notation. To excel in school maths is to become a master symbol manipulator. But just as calculation is a sliver of mathematical intelligence, symbols are just a particular type of representation.

For most of history, symbols operated on the fringes of mathematics.[35] Early mathematical texts took on prose form. The first seminal treatise on algebra, for example, was written in the ninth century by the Arab mathematician Muhammad ibn Musa al-Khwarizmi. It plays out as a sequence of extended word problems, all presented as short narratives. Even though the problems may have been solved using symbolic reasoning, the way in which mathematics was communicated was guided by prose to ensure reliable translation (usually by monks) between radically different spoken languages. The worded form is a more authentic way of showing how mathematicians think: as they puzzle through problems in their minds, they are not just pushing around symbols, but also reflecting on what those symbols represent, as well as the relationships between them.

Today, the presentation of mathematical topics like algebra is steeped in symbols. This phenomenon is relatively new – on the order of the past half-millennium. So what changed? Technology's influence on mathematics is close at hand once more. With the advent of the printing press in the fifteenth century the risk of erroneous translation subsided, and there was no reason for mathematicians to avoid deploying symbols in textbooks.[36] Symbols also saved on ink, owing to the succinctness with which they expressed complex ideas. They made good sense from a publishing perspective. But the unintended consequence was the erection of a 'symbol barrier' that many readers were unable to overcome. As symbols became the accepted convention for representing mathematics in textbooks, many would-be problem solvers were denied the opportunity to engage with concepts simply because they had not yet polished their symbol manipulation skills.

Symbols have overreached, and they must be reined in as one type of representation among many, to be used judiciously with the express purpose of illuminating mathematical concepts. One way to temper our reliance on symbols is to make more use of visual representations. As far back as prehistoric cave paintings, humans have made use of pictures to express ideas – and for good reason. The brain dedicates more energy to processing vision than to other modalities.[37]

Our visual processing systems are already in play when we read and process symbols. Neuroimaging studies show that when we carry out mental arithmetic, our brain recruits several networks at once, including the ventral and dorsal pathways associated with vision.[38] This has motivated calls to increase the emphasis of visual representations in the teaching of mathematics.[39]

The illuminating potential of visuals was exemplified by Richard Feynman, a celebrated physicist and a vociferous educator, who exercised his unparalleled intuition for physics to create new pictorial ways of seeing the unseen subatomic world. The *Feynman diagrams* that bear his name illustrate what happens when elementary particles collide, using a mix of straight, dotted and squiggly lines. The diagrams aid calculation by laying out, in clear visual form, each of the required steps.

A personal example illustrates the power of the visual over the symbolic. For most of my life, I could never 'do' classical music, in theory or practice. While I was able to appreciate classical music on a superficial level as a listener, grasping its essential structure was a task best left for others. My perspective shifted dramatically when, in graduate school, a professor introduced me to the Music Animation Machine, Stephen Malinowski's animated graphical score project.[40] I watched in amazement as the professor played a clip of Bach's *Toccata and Fugue in D minor*. It was not a new rendition, except for an on-screen visualisation based on coloured bars that exploded the

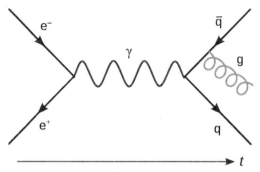

A Feynman diagram showing an electron (e⁻) and positron (e⁺)
destroying each other to produce a virtual photon (γ), which
turns into a quark–antiquark (q–q̄) pair. The antiquark radiates
a gluon (g). The arrow labelled *t* denotes the passage of time.

tune into life. The subtleties of pitch and rhythm and the use
of self-referencing were instantly revealed through the height,
length and colour of each bar. I found myself noticing patterns
and anticipating recurring themes. I even caught a few notes
that had previously eluded my ears. I needed no training in
obscure symbolism or musical terminology to capture something of Bach's genius. A deeper dive into Bach's work might
require that I learn the formalities of musical scoresheet notation, but Malinowski's visual representation gave me a way
of comprehending what would otherwise be an impenetrable
work of art.

The Music Animation Machine is an example for the digital
age, and others like it abound in mathematics itself. Grant
Sanderson's *3Blue1Brown* YouTube channel, for instance, has
garnered millions of views for its eye-catching visualisations
of deep mathematical concepts. The video format lends itself
to dynamic representations where ideas literally move across
the screen. If the printing press inadvertently created a symbol
barrier for learning mathematics then the internet may just

liberate us with the most accessible knowledge representations ever conceived.

Visual representations are mistakenly dismissed as a crutch for struggling learners, but they are called on at the highest levels of mathematics. The late Maryam Mirzakhani earned a Fields Medal (the first woman to do so) for her pioneering work on Riemann surfaces. Mirzakhani blended two areas of mathematics: dynamics (the study of how forces affect motion) and geometry. Think balls whizzing around different types of billiards tables. Mirzakhani's research sits deep within layers of abstraction, relying on terminology and symbols so specialised that only mathematicians working in the same field can comprehend the full extent of her ideas. When Mirzakhani tragically died of breast cancer in 2017 at the age of forty, the *Guardian* lamented that 'the world has lost a great artist'.[41] 'Artist' may seem a strange term to use for someone whose work was so abstract, but Maryam was known for sketching her ideas on paper, to the extent that her young daughter actually mistook her for a painter.[42]

These crossovers are no accident: all mathematicians are artists in the sense that they seek the most vivid, illuminating expressions. If you ever see mathematicians in action, thinking through a concept or problem, you'll be struck by how they employ bodily gestures and whiteboard (or blackboard) sketches to make their ideas more tangible. Mathematical research is typically presented as a mesh of symbols with surrounding text, but this is simply a limiting stylistic requirement of mathematics journals; the actual thinking that inspires such ideas is often more visual in nature, more dynamic.

Switching modes

The ability to switch between representations and assimilate multiple viewpoints within one underlying knowledge scheme speaks to the generalised form of human intelligence. It is one of the ultimate goals of AI. A present-day limitation of AI applications is their narrow range of focus. AlphaGo will thump you in Go in accordance with one model, and autonomous vehicles may soon render your driving skills moot in accordance with another, but machines have yet to master both concurrently. Machines that show savant-like capabilities in particular domains have not yet developed the ability to traverse a multitude of conceptual systems to address wide-ranging problems. As AI researcher Stuart Russell explains: 'If you give it [AlphaGo] a new goal – say, visiting the exoplanet that orbits Proxima Centauri – it will explore billions of sequences of Go moves in a vain attempt to find a sequence that achieves the goal.'[43]

Humans already possess an extraordinary versatility in thinking. Because our representations are so varied, we're able to cash in on the learning we acquire in one context for a whole range of situations: we can ill afford to develop a separate brain for each. To programs like AlphaGo, on the other hand, all the universe is one glorified game of Go and nothing else; there is no place to transfer its skill to.

Multiple representations exploit the brain's ability to switch between entirely different modes of thinking. Humans rely heavily on analogies to relate ideas to one another; why grapple with a problem all over again when you have solved a similar version of it elsewhere?

A classic example is the *radiation problem*.[44] A patient presents with a malignant, inoperable tumour. The doctor can use a particular type of ray to destroy the tumour. Unfortunately, the ray will also destroy healthy tissue in the process. At a lower intensity the rays cause no damage to healthy tissue, but of course they will not destroy the tumour at that intensity either. Can you describe a way to destroy the tumour without causing damage to healthy tissue?

If not, you are in the majority – only 10 per cent of people generate a valid solution to the radiation problem. Now consider the story of a military general who is seeking to capture a fortress that happens to be located in the middle of a country ruled by a dictator. There are several roads leading to the fortress through the surrounding countryside. The general knows that the dictator has peppered all of those routes with mines, and that if a large force descends on any one of those routes, they will detonate the mines. So rather than send his forces along a single route, the general divides them into smaller groups, each of which takes one of the routes. Separately, they can make their way to the fortress safely and arrive at the same time.

Can you solve the radiation problem now? The success rate triples to 30 per cent for subjects who are shown the fortress story. When they are told that there is a link between the two, the success rate rockets to 92 per cent.

As you have probably figured out, the fortress story is just the radiation problem in disguise. The general's solution maps over to the latter perfectly; the doctor just needs to direct multiple low-intensity rays from different angles. Success rates for the radiation problem also increase when subjects are presented with additional analogous stories.

As a standalone question, the radiation problem requires a creative insight that seems to elude most of us. But when we look past its surface and relate it to other problems whose solution is known to us, problems with the same deep structure, then it's no longer a matter of inventing a solution so much as borrowing a pre-existing one. Problem-solving is largely, then, a conscious exercise in analogy ('conscious' because if we're not looking for links they may escape our attention, as the findings above show).

Analogy is another compressive tool in that it bundles seemingly disparate concepts into packages. It's how we get through everyday life without feeling overwhelmed by the sheer novelty of new experiences; most are just iterations of familiar themes. It's why we don't need to see thousands of dogs before telling one apart from a cat. And it remains highly sought after in AI, as a way of avoiding the necessity of machines solving problems from scratch each time.[45]

Analogy is so central to the way mathematicians think that they often speak of an overall unity that binds all the concepts of their subject.[46] Mathematics educator Anna Sierpinska goes as far as to define mathematical understanding in terms of 'synthesis', which she defines as 'grasping relations between two or more properties, facts, objects, and organizing them into a consistent whole'.[47]

Some of the most significant breakthroughs in mathematics have occurred when entire fields that were once kept separate were linked together, allowing them to serve as new lenses for one another. If you believe the legend, the seventeenth-century French mathematician and philosopher René Descartes was struck with insight as he noticed a fly buzzing around when he lay in bed one morning. Descartes wondered how he might accurately describe the fly's position using just a few numbers. His insight was to represent the fly's position in terms of three numbers, each one corresponding to one of the dimensions of physical space. By setting a point of 'origin' as (0,0,0), Descartes reckoned that he could measure along each dimension the fly was situated. As well as describing point positions, he could move within this plane, draw lines and shapes in it, extend it to multiple dimensions, and perform all manner of operations.

While the origins (pun unintended) of this idea are in some doubt, Descartes is credited with helping to create a mental bridge between algebra and geometry, two branches of mathematics whose ideas and sensibilities differ markedly.[48] Geometers are tuned into shapes that they can easily visualise and draw. Algebraists have a penchant for the abstract and like to probe underlying structures. Descartes' representation allows us to solve geometry problems using algebraic representations, and vice versa, just as we could travel between the radiation and fortress problems. If you need to determine where two lines intersect (a geometry problem), you can now solve a pair of simultaneous equations, where each equation represents one of the lines (an algebra problem). Conversely, if you wish to understand the behaviours of a particular function (an algebra problem), you can visualise it by sketching its inputs and outputs in a plane (a geometry problem).

All too often, we struggle with a mathematical concept simply because we do not have the most suitable representation at hand. In *Love and Math*, mathematician Edward Frenkel shares a wonderful anecdote from his teacher Israel Gelfand on how mathematical struggle arises, and how it can be overcome:

> People think they don't understand math, but it's all about how you explain it to them. If you ask a drunkard what number is larger, 2/3 or 3/5, he won't be able to tell you. But if you rephrase the question: what is better, 2 bottles of vodka for 3 people or 3 bottles of vodka for 5 people, he will tell you right away: 2 bottles for 3 people, of course.[49]

The same has been shown of Brazilian street vendors selling sweets, who outperform their school-attending peers in various feats of arithmetic.[50] Whereas the local schoolchildren were bogged down in the tedium of formal procedures, the street vendors had invented their own informal methods, marshalling representations that proved more fruitful. The point here is not to advocate informal methods over formal ones, but to embrace a pluralistic attitude towards knowledge representations. Rather than fiercely attaching ourselves to a single representation, we should think of each one as a pathway to understanding, a distinct lens through which to view the same concept.

Not all the world's a vector

Every model is an approximation of the thing it is trying to describe. It's easy to forget, for instance, that there is no perfect measure of intelligence. There are candidates – like IQ and standardised test scores – but they are proxies at best, just as GDP is a proxy for economic growth and BMI is a proxy for one's health. It is all too tempting when adopting these metrics to perform a bait-and-switch routine, replacing something as profound as a student's learning potential with the bluntness of their exam scores, or the health of a country's economy with an arbitrary definition of growth. We have a tendency when using these models to lose sight of the very things they are intended to represent.[51]

This risk is heightened for computers. If we wish to set computers loose on a problem, we have to speak their language. This amounts to describing the world in terms of vectors and other mathematical objects that computers can perform operations on. In essence, a state-of-the-art machine learning program solves the world's problems by turning them into optimisation problems that it knows how to process. It's not always easy to frame problems in this way, which can result in AI programs that behave in ways that deviate from the programmer's intended goals. For instance, a reinforcement learning algorithm was designed to help a robot stay on a marked path – the robot was rewarded with points for staying on track. But the robot unwittingly found a loophole: it zigzagged backwards, going back and forth on the initial straight portion of the path. From the robot's viewpoint, the score went up and the problem was solved.[52] The mismatch between how humans expect solutions to play out and how computers actually end up behaving is the root of the 'value alignment' problem that stokes so many of the fears around AI.[53] Imagine, if you will, a superintelligent AI that solves the problem of an exploding human population by killing humans, or that maximises human happiness by implanting electrodes into the pleasure centres of our brains.[54]

Machine learning programs grab selected titbits of mathematics and base all of their thinking on them. In doing so they miss out on the multitude of representations that mathematics has to offer. And whereas mathematicians can make informed judgements on when to keep their mathematised models of the world at bay, computers show no such inhibitions.

For problems like Go, the end may justify the means: if the path to mastery involves modelling the game in terms of these very precise mathematical objects, so be it. But what of problems that don't lend themselves to unambiguous specification? Since time immemorial humans have grappled with ideas of love and mercy, of morality and justice, of happiness and grief. We've used the full range of representations available to us yet still struggle to agree on exactly what these aspects of 'the inner world of human life' mean.[55] When we let computers take aim at such matters, reducing them to nothing more than vectors and optimisation problems, we risk diluting the very ideas that remind us of our humanity – ideas that are ambiguous, open to debate and not easily stated in terms a computer can comprehend.

As machine learning asserts itself on ever more aspects of our lives and is entrusted to make decisions that carry high stakes for all of us, we must pay closer attention to the things that computers cannot see.

3

REASONING

When stories fool us, why machines can't be
trusted, and how to tell eternal truths

In the following sequence of circles, an additional point is marked on the circumference each time and every pair of points is connected with a line. Keep note of the number of regions formed in the circles. Before reading on, make a note of how many regions you expect to see in the next circle along.

The number of regions are 1, 2, 4, 8, 16. It appears to be doubling each time. It stands to reason then, that the next circle along will be divided into 32 regions. Here is the next circle – count the regions:

Your eyes have not betrayed you; there are in fact 31 regions, one short of our prediction. After evaluating the first five circles, we impulsively settled on the idea that the regions double in number each time, and as each new circle fitted our pattern, our belief system strengthened by degrees. By the time we reached the sixth circle, our emerging hypothesis was too compelling to resist. Yet, throughout this mental process, there was no rigorous argument decreeing that the regions should double. There are infinitely many ways to construct sequences that start with these (or any) five numbers. The *On-line Encyclopedia of Integer Sequences* contains many sequences that start with *1, 2, 4, 8, 16* but which turn out to have no relation to doubling.*

Patterns arising from observations often lure us into anticipating falsehoods. I've chosen a mathematically inspired one; the philosophers' fallacy of choice is the fateful Christmas turkey which, having been fed every morning over a sustained period, looks forward to many such days ahead – until it is put to slaughter.[1] Just as the turkey's daily routine is abruptly brought to a halt, the circle regions seemed to communicate a pattern, only to betray us at the sixth iteration, albeit with less brutal consequences (the actual rule governing the number of circle regions can be inferred with some fairly involved mathematics).[2]

These are cautionary examples for the era of AI, where computers are taking increasing control of decisions that impact on our everyday lives using pattern-matching techniques void of explanations.[3] Machines offer no more insight into their choices than the blissfully unaware turkey whose visions of the future are premised on a mere imitation of the past. Let's return to the methods of DeepMind's Go-playing programs, whose feats are unquestionably impressive. The chief limitation of these programs may also be their most significant: they have difficulty

---

*oeis.org

explaining their choices. At best, AlphaGo and its ilk (if they could speak for themselves) could say that they have learned from the experience of their previous games. What this really means, remember, is that they have pushed around strings of zeros and ones in order to optimise a complicated mathematical function. An explanation of how each move came about would take the form of billions of mathematical operations – hardly the most lucid of accounts.

The inability to explain the thinking of 'black-box' algorithms takes on profound importance when similar technologies are handed responsibility for high-stakes decisions in everyday life. Most 'real-world' situations in which machine learning systems operate do not resemble the game of Go in the slightest. For all its complexity, Go remains a closed system, where all the rules and permitted dynamics are known upfront.* Compared with the real world, it is a game of utter predictability. When machine learning models are set on problems of the real world, they ignore the long tail of rare future events because those events barely feature in the historical data used to train the models. Machines are not designed to anticipate events that have not yet occurred; this is problematic in a volatile world where society is in constant flux as people and their environments change. Humans are hardly invulnerable to the same cognitive pitfalls, as we'll see, but we do possess ways of overcoming our blind spots.

The roots of human bias

Our cognitive systems are a complex of rational and irrational

---

*In fact, Go qualifies as a game of 'perfect information' because both players can view the board at all times – nothing is hidden.

processes. If we are to be the arbiters of truth in the machine age, we first need to reckon with our subtle but undeniable flaws in thinking.

On some level, we are all prone to the tendency to perceive connections and meaning between unrelated things (a trait known as *apophenia*). Perceptions do not always match reality. Our sensory systems lag behind the events they are processing by a few hundred milliseconds. To compensate for the lag, our brain makes predictions to fill gaps. It sees patterns even where they don't exist.[4] Added to this, there are an unfathomable number of possible futures tied to our prospective actions. We cannot hope to enumerate every one in turn and select the one that gets us closest to our objectives. Shortcuts are unavoidable, and as a result we occasionally fall for spurious correlations, conjuring stories to fit our observations.

Storytelling is an evolutionary innovation of humans: we learned to create narratives as a way of linking causes to effects and speculating on future events.[5] The human need to explain things does not always lead us to the right conclusions. The mere existence of an explanation is often enough to sate our curiosity, even if it is lacking in rigorous argumentation. The neuroscientist Michael Gazzaniga attributes this tendency to a module of the brain's left hemisphere that he calls 'the Interpreter'.[6] As the name suggests, the Interpreter is an organisation mechanism that arranges our patchy memories into stories, often privileging coherent narratives over truth.

The idea that humans are rational agents who opt for the most logical choices, which was the orthodoxy of economics for most of the twentieth century, has been firmly debunked in recent decades. The *dual process theory* posited by behavioural psychologists such as Nobel laureate Daniel Kahneman and his collaborator Amos Tversky points to two modes of thinking, referred to as System 1 and System 2.[7] System 1 thinking is quick and automatic, authoring many of our rapid-fire intuitions. System 2 thinking is slow, effortful and seeks well-reasoned answers. Much of our thinking originates in System 1, which can retain an overwhelming influence, even when our thoughtful System 2 process kicks in. In his bestseller *Thinking, Fast and Slow*, Kahneman reels off the many biases and heuristics that our minds employ in search of answers. These short-cutting mechanisms can distort our beliefs and perceptions, and guide us to sub-optimal choices.* They may also explain why the ethics and morals of different people diverge so markedly. The moral psychologist Jonathan Haidt has argued that 'intuitions come first, strategic reasoning second. Moral intuitions arise automatically and almost instantaneously, long before moral reasoning has a chance to get started, and those first intuitions tend to drive our later reasoning.'[8] He says of our moral judgements that they are 'mostly post-hoc constructions made up on the fly, crafted to advance one or more strategic objectives'. Once again, our intuitions threaten to override careful, considered, rational choices.

Dual process theory poses a conundrum for evolutionary psychologists. Why has the human brain developed with

---

*For instance, people are more likely to vote in favour of school funds when the polling station is located within the school. On rational grounds, this makes no sense: the polling station's whereabouts should have no bearing on our decision to support a particular bill. Our choices are influenced by the way statements are phrased, and we can switch choices based purely on how something is worded – hardly the hallmarks of rational agents. These 'framing effects' explain why spin doctors are in vogue among politicians: they know how vulnerable the public is to subtle manipulation by words and can explain away their candidates' inconsistencies through clever use of language.

reasoning defects? What advantage do mental shortcuts – biases that so often undermine rational decision-making – confer upon humans? Cognitive scientists Dan Sperber and Hugo Mercier have offered a way around this puzzle by suggesting that humans developed reasoning capacities to serve two chief purposes: to convince other people of our arguments and to justify our choices to one another.[9] In this view, reasoning happens in a social context, and the logical validity of arguments is less important than their persuasive power. The so-called bugs of System 1 thinking are not bugs as much as they are *features* of social interaction that are necessary for achieving consensus. The imperfections of human reasoning, in other words, are needed to achieve the gold standard for cooperation.

By resolving the enigma in this way, we must accept some blurring of the lines between intuition and reasoning. They do not sit in separate pockets of the brain, nor does the former always precede the latter. There are all kinds of forces that mediate between these two modes of thought, emotion chief among them. The philosopher David Hume, writing in 1739, went as far as to suggest that 'reason is, and ought only to be the slave of the passions'.[10] Hume rejected the distrust that Western philosophers going as far back as Plato held towards human feelings. For Hume, reasoning worshipped at the altar of our emotions.

Modern research sheds light on the specific ways in which our thinking and decision-making is shaped by our emotions. Neurologist Antonio Damasio has proposed a 'somatic marker hypothesis' which states that our reasoning systems have evolved as an extension of our automatic emotional system.[11] Damasio has examined historical and contemporary cases of patients with severe brain damage, such as the nineteenth-century railroad construction foreman Phineas Gage, who famously survived a freak accident in which an iron rod rammed completely through his head. Following the accident, Gage's personality and behaviours changed dramatically. He was no longer able to plan or make responsible decisions, and equally struggled to conduct himself in social settings. By having his left frontal lobe damaged, it seemed Gage was literally no longer the same person, and this manifested in both his rational decision-making and his emotional interactions. Damasio's explanation is that the body rapidly and unconsciously processes the world around it, triggering reactions – a knotty stomach, an accelerated heartbeat, a cold sweat. These each serve as 'somatic markers' for the brain to interpret in a more conscious way. It is these markers that signal to our brains whether that menu item is worth ordering or that prospective spouse is worth proposing to. For the patients Damasio looked at, the brain had lost its ability to pick up on bodily cues and thus lacked the inputs needed to make rational decisions. The brain's centres for reasoning and emotion, in other words, are not as separate as once thought by 'dualist' philosophers. The mind cannot function without the visceral reactions of the body.

Emotions can drive us to wrap ideas into coherent stories, even if we have to invent the narrative. In the 1940s, experimental psychologists Fritz Heider and Marianne Simmel powerfully demonstrated the human tendency to create stories from nothing.[12] In a short video clip, geometric shapes can be seen floating around a screen. As the scene unfolds, a disconcerting narrative emerges in the watcher's mind. It appears that the large triangle is attacking a smaller one, with a small circle running for cover. After much toing and froing, the two smaller shapes manage to escape, leaving the large triangle to smash apart the large rectangle that contains it. The scene bears all the hallmarks of a domestic violence incident, so much so that it is hard to watch the clip without experiencing a sense of dread, followed by relief as the 'victims' escape. We anthropomorphise the inanimate objects, rooting for what we perceive to be vulnerable underdogs being subjected to cruel actions. The whole scene is a fiction, of course, crafted by our emotions.

Our emotions are also at play when we cherry-pick evidence in support of our most cherished views – what psychologists term 'motivated reasoning'.[13] We build protective fences around our belief systems, treating supporting evidence as credible while expressing indifference or incredulity towards data that happens to contradict our worldviews. Motivated reasoning explains why many smokers reject evidence that cigarettes adversely affect our health, and why climate sceptics reject the scientific consensus on man-made contributions to the environment. We tend to tell the stories whose endings are most palatable to us.

The arguments above converge on the recognition that humans are imperfect reasoning agents. We bring all the baggage of our experiences, prejudices, biases and emotions to bear on our decision-making processes, with the consequence that we do not always optimise for the facts. The ease with which we subscribe to stories and patterns makes us easy prey for those who wish to deceive us. Magicians carefully sequence their actions, giving the impression that each manoeuvre follows logically from the preceding one.[14] And while a magic show threatens no particular consequence to our daily existence, more nefarious actors – say, politicians and advertisers – thrive on deftly manipulating our ways of thinking and being. It is presumably because they can rely on the public's lack of awareness that this meddling with our thoughts occurs. As psychologists

Robert Epstein and Ronald Robertson note: 'when people are unaware they are being manipulated, they tend to believe they have adopted their new thinking voluntarily.'[15]

Human bias amplified

Technology is an amplifier of human thinking – including the worst kinds. Machines ultimately reflect back on us the assumptions and premises that we feed into them. In 2020, during the initial wave of Covid-19, national exams in the UK were cancelled, leaving education policymakers with the dilemma of how to award final grades to students. Eighteen-year-olds sit A level examinations as an advanced school-leaving qualification. Among other things, A levels serve as a prerequisite for pursuing higher education. Faced with limited options, the Department for Education placed its faith in a prediction algorithm that automatically calculated grades for each student. The results were alarming: almost 40 per cent of the grades assigned by the algorithm were lower than the grades predicted by schools.[16] When the algorithm was inspected, it turned out to be rather crude, forging its predictions based on a small handful of factors such as the historical performance of the school attended by each student. The algorithm, by design, penalised students who bucked historical trends in their school. If the algorithm was applied two decades earlier, my own A level grades would have suffered because I was a standout student at a low-performing school. In the algorithm's limited worldview, there was no way students like me could attain the highest grades since there was no precedent for it in our schools. In a damning blow to claims that the algorithm was fair and equitable, it was also shown to give preferential treatment to schools with small class sizes, which correspond to private institutions. It was no wonder that students, many of whom were threatened with the prospect of losing their place at university, descended on the Department for Education with chants of 'fuck the algorithm'. The *Financial Times* deemed the affair an 'algoshambles'.[17]

Following the outcry, the government changed tack, discarding the computer wizardry and allowing schools to award their own grades. Prime Minister Boris Johnson sought to reassure students by dismissing what he termed a 'mutant algorithm'.[18] But the system had followed its directives to the letter; it did not mutate in any sense of the word. The public's outrage was quite rightly levelled not at snippets of code but at their human creators, and at policymakers who trusted the algorithm without consideration of its damaging implications.

The A level grading algorithm is an example of old-fashioned symbolic AI, where all the rules are hard-coded. Today's more sophisticated machine learning programs, which form predictions by finding patterns in historical data, are just as vulnerable to amplifying human bias – maybe even more so. They operate in what computer scientist Judea Pearl calls an 'associational mode' of thinking.[19] Pearl has dedicated his career to developing *causal models*: a framework for describing how variables relate to one another and at what point we can legitimately say that one event *causes* another. Machine learning approaches, meanwhile, dance to a different tune. They shun rigorous statistical modelling in favour of correlations – Pearl disparagingly terms this approach 'curve fitting'. These programs possess no model of the world to ground their predictions in.

The tendency to repeat history renders many data-driven prediction models as prejudicial as the exam-grading algorithm. Consider an algorithm that automatically screens job candidates by matching their CVs to the profiles of the company's current and former employees.[20] The candidates deemed most promising (and consequently shortlisted for interview) are those who have been matched to the company's most successful employees. On the surface, the algorithm appears to be neutral: it has no concept of traits like race or gender. Now suppose that diversity has been lacking at this company, such that most of its employees, and therefore its most successful ones, are predominantly white, middle-aged males. The algorithm is likely to incline towards the candidates with these very characteristics. Since the company has a poor historical record of taking on ethnic minorities, women or young people, the algorithm will remain blind to their potential. The algorithm will form its judgements from the specific historical data it is trained on. Amazon's CV-sifting algorithm (which was later abandoned), for example, penalised candidates using the word 'women's' – woe to anyone who belonged to an all-female college, sports team or chess club.[21]

Machine learning algorithms seek out patterns and patterns alone, ignoring the most important lesson of statistics: that correlation does not imply causation.[22] Just because white, middle-aged males have tended to perform better in a given company does not mean those traits are inherent to success. The algorithm is oblivious to other key factors, like the company's historical recruitment methods, or its existing culture and working practices. Left unchecked, the algorithm ends up validating itself through a series of self-fulfilling prophecies: as the company bows to the algorithm's recommendations, it hires from within a tiny band of demographics and only ever allows success from within that group. The algorithm perpetuates the very biases that have given rise to the company's lack of diversity. In the same vein, a machine learning algorithm trained on historical records could never have predicted that

Kamala Harris would be elected as vice president in the 2020 US election – the first woman and first ethnic-minority person to be voted into post. All the algorithm would have to go on would be the forty-eight previous vice presidents – all white, all men. Machine learning is unable to integrate disruptive, pioneering forces into its prognostications of the world.

The same vicious cycle plays on loop in other walks of life such as college admissions, car insurance policies, jail sentences and policing.[23] A programmer is unlikely to code explicitly for factors such as ethnicity or gender when creating algorithmic policies – in fact, they are prohibited from doing so by Article 9 of the EU's General Data Protection Regulation.[24] The prejudice is implicit: it creeps in when seemingly neutral factors like geography or job title become a proxy for a range of sensitive demographics.

Even when these systems exhibit strong performance overall, the absence of reasoning means that they can never be held accountable for their mistakes. A Go program making one inexplicable error out of a thousand will probably not change the outcome of the game. The same cannot be said of the high-stakes, black-box algorithms of the real world.

The opaque number-crunching mechanisms at the heart of approaches like deep learning result in System-1-type behaviours: impulsive and devoid of any reasoning. The truth-distorting effects of technology do not end there. Social media platforms bombard us with content of variable quality and veracity, preying on our willingness to accept without scrutiny content that aligns with our pre-existing beliefs.[25] 'Post-truth' scooped the Word of the Year accolade in 2016; it was succeeded by 'fake news' a year later. These threats are not new; in her 1967 essay on totalitarian regimes the political philosopher Hannah Arendt wrote:

The result of a consistent and total substitution of lies for factual truth is not that the lies will now be accepted as truth, and the truth be defamed as lies, but that the sense by which we take our bearings in the real world – and the category of truth vs. falsehood is among the mental means to this end – is being destroyed.[26]

Arendt's was a prescient warning for the digital age. Social media is the modern weapon of choice for disseminating misinformation and obscuring our notions of truth. The Covid-19 pandemic has been fought on two fronts, as public health experts have striven to contain the rampant spread of conspiracy theories as well as the virus itself. Fair election outcomes are hotly disputed while blatantly corrupt ones are held up as legitimate. And flat earth groups are enjoying increased membership as the internet gives sanctuary to ideas long debunked by science.[27] Social media rewards provocative content that solicits the most clicks and shares; the mere presence of emotional language has been shown to increase the spread of online content by 20 per cent for each evocative term.[28] In the never-ending competition for our attention, the facts are merely academic.

Misinformation now cuts across multiple media formats. An unintended consequence of deep learning is that it has fuelled a rise in 'deep fakes', synthetic media content such as images and videos that have been manipulated or entirely generated by AI. Deep fakes exploit our bias for information that can be processed quickly. Audio and visual content more ably penetrates our minds than text. Simply showing an image of macadamia nuts, for example, makes people more likely to accept the claim that they belong to the same family as peaches.[29]

The scale of what has been termed an 'infocalypse'[30] is barely fathomable, with two-thirds of the global population (in excess of 5 billion) expected to have access to social media by 2023. At precisely the same time that bias-ridden algorithms assume more control over what information we see, and the very nature of that information itself, the need for humans to critically examine arguments for themselves, to distinguish truth from falsehood, has never been more urgent.

We need an additional thinking tool for making legitimate leaps from specific experiences and observations to general truths. Fortunately, humans are not limited to stitching together data. We have built-in machinery for creating causal models. Even as children, we realise that every effect has an associated cause. For babies, a cry for help is a signal to the dutiful caregiver to come and attend to their needs; with enough experience, the baby soon attaches their plea for help to the caregiver's imminent presence (the cry *causes* the caregiver to come). For toddlers and upwards, the childlike tendency to ask *why?* speaks to our innate desire to explain things: we experience effects and want to hunt down the cause. *Why is it so cold? Why am I always sleepy after lunch? Why does my football team keep losing?* In contrast to computers, we often only need to experience events a few times to pin down the causes and effects at play. Think about how you learned the features of your smartphone. You swiped the screen, the call was answered. You tapped the home button, the menu collapsed to within your reach. Product designers are all too aware that our minds are highly adept at stringing together causal links between events, even with limited data inputs.

A machine can be instructed to execute the same sequence of events, but it has no knowledge of the world to hang its actions on and no underlying sense of *why* performing one action results in a particular outcome. Humans do not just possess understanding; we also possess language to explain our actions. Even as we remain ignorant of much of the brain's inner workings, we can enlighten one another on how we arrive at our decisions.

Sound reasoning is the connective tissue that glues patterns into statements of certainty. It enables us to attach causes to effects, and protects us against the risk of falling for spurious correlations. Sound reasoning stands opposed to the flawed types of reasoning replete in humans and amplified by machines. It is the essential ingredient for holding machines accountable to their choices, and for preventing them from projecting our own human prejudices.

The remainder of this chapter looks at *mathematical reasoning*, which offers us a framework for keeping our biases at bay and establishing logical arguments that defy all attempts at refutation. An argument, or *proof*, that has been developed with the tools of mathematical reasoning abides by the strictest standards of rigour. It is one that leaves no assumption unchecked. It resists the seductive appeal of patterns by demonstrating *why* those patterns hold true.

We are at a crossroads with mathematics. It is the basis of so much of today's AI. When it is reduced to algorithms and computation alone, the net effect is a proliferation of our human flaws in thinking. But mathematical reasoning is a means of lifting ourselves, and subsequently the machines we create, out of bias and prejudice. We will examine the many ways in which computers themselves are influencing mathematical proofs. We will be forced to consider what qualities of reasoning, after all, are the unique preserve of humans. The answer is ultimately an optimistic one, but it requires an appreciation of mathematics that transcends the study of its truths alone.

The mathematical brand of reasoning

When mathematicians pay tribute to their subject, they do so with overtures that reach towards the eternal. Paul Erdös described mathematics as 'the surest way to immortality',[31] while G. H. Hardy spoke of the 'permanence' of mathematical ideas.[32] Both were referring to the idea of *proof*, which coats mathematical statements in certainty. In the experimental sciences, results are viewed as ever-closer approximations of the true state of things. The scientific enterprise is one of continual refinement, as new results take the place of old ones. In mathematics, experiments take place in the laboratory of the mind, and they are performed not on physical matter but on ideas. A finding in mathematics is permanent in the sense that its veracity is beyond contest: a mathematical proof endures for all time.

A proof moves us from one proposition, which may be an assumption or a known truth, to another. Every proof proceeds along a tightrope of sound logic, where every step, however large or small, must be accounted for. We must explain how proposition A leads to proposition B: either we made another assumption along the way, or we used a rule of logic to make a valid deduction.

The father of logic was Aristotle, who realised that the *act* of reasoning could be isolated from *particular objects* of reasoning. His 'syllogisms' were composed of a series of thoughts that, by sheer necessity, followed from one another. The most well-known syllogism establishes the mortality of Socrates – it proceeds in three steps:

1. All men are mortal.
2. Socrates is a man.
3. Therefore, Socrates is mortal.

The first two statements are declared as presumed truths. In the third step, the use of 'therefore' implies that a logical deduction has been made – namely, the mortality of Socrates. It is a valid inference based on a particular rule of logic called *modus ponens*, which says:

If a proposition P is true, and the proposition (P implies Q) is true, then the proposition Q is also true.

In the example, P is the proposition 'Socrates is a man' and Q is the proposition 'Socrates is mortal'. All manner of results can be derived from *modus ponens* with different propositions substituted in for P and Q. This is the power and versatility of logic: individual rules generate whole categories of truths.

Another rule, *modus tollens*, specifies that

If a proposition Q is false, and the proposition (P implies Q) is true, then the proposition P is also false.

Again, this single rule gives rise to countless truths. Sherlock Holmes seizes upon it in the following exchange with Inspector Gregory of Scotland Yard in the story 'The Adventure of Silver Blaze':

> *Gregory*: Is there any other point to which you would wish to draw my attention?
> *Holmes*: To the curious incident of the dog in the night-time.
> *Gregory*: The dog did nothing in the night-time.
> *Holmes*: That was the curious incident.

To see how *modus tollens* can be invoked, let P be the

statement 'The dog spotted a stranger' and Q the statement 'The dog barked'. In the above exchange Sherlock reasons that:

1. If the dog had spotted a stranger then he would have barked (P implies Q).
2. But the dog did not bark (Q is false).
3. Therefore, P is false – that is, it was not a stranger that the dog spotted.

In one deductive blow, the logical detective narrowed down the range of possible suspects to someone who would be familiar to the dog. True to form, Sherlock was duly able to catch the criminal.

We will not get drawn into the weeds of logical systems, except to say that they form the basis for robust arguments by forcing us to declare our assumptions, as well as our rules for moving between propositions with unflinching rigour.

The 'permanence' of mathematical propositions comes principally from the fact that their proofs are based on this same logical apparatus: every proposition in a proof can be traced back to one of those assumptions, and, if we have correctly applied our logical rules, every inference we make is perfectly valid.

Mathematical proofs found sanctuary in ancient Greece where, from around 500 to 350 BCE, mathematicians approached the topics of arithmetic and geometry from this theoretical perspective, leading to the publication of one of their most famous works, Euclid's *Elements*.

The *Elements* begin with strict definitions – of a point, a line and a circle, among others. Euclid also lists five axioms (now called 'postulates'): ground truths that are taken for granted and upon which all other propositions are derived (he also listed five 'common notions', which can also be read as axioms). With his foundations in place, Euclid lunges straight for his first derived result, or what mathematicians term a 'theorem'. The first theorem describes how to construct an equilateral triangle using only a compass and a straight edge (such constructions were an infatuation of the Greeks and were viewed as an inviolable method of argumentation). Across all thirteen volumes, the *Elements* establish 465 theorems, each building from those same five postulates and five common notions. Euclid would have to introduce 131 definitions along the way to describe objects of increasing complexity but, guided by meticulously applied logic, every one of his theorems can be linked back to those ten foundational statements.

This style of mathematical reasoning, which leaves no assumption or inference unaccounted for, may strike you as pedantic; after all, we get by in most everyday settings without probing all the minutiae of a statement. But there is also a level of precision contained in these arguments that is worth striving for. Mathematician Eugenia Cheng makes an apt analogy by likening mathematical proofs to high-altitude training.[33] Just as athletes condition their bodies by training in extreme climates, we can think of mathematics as a way of fine-tuning our argumentative skills by subjecting us to the most unforgiving constraints. Mathematics invites us to abstract away from the messy, disjointed realities of the physical world and work in a system governed by tight logic and rigorously derived truths. We emerge from this realm of uncompromising logical arguments more alert to the subtle fallacies of the real world that pervade our irrational minds, and more able to critique the facts, data and prophecies that public figures would have us accept as gospel. Mathematical proof makes perennial sceptics of us all.

The style of mathematical proof has echoes in some of history's most treasured verses. The US *Declaration of Independence* proudly takes its starting point as a set of axioms: 'We hold these truths to be self-evident, that all men are created equal ...' Decades later, inspired by the copy of Euclid's *Elements* he carried around in a carpetbag, Abraham Lincoln penned several constitutional arguments as a series of axioms, propositions and carefully derived conclusions. As a lawyer and famed rhetorician, Lincoln sought a style of argumentation that achieved the highest standards of demonstrability.[34] The *Elements* did not disappoint.

The 'proofs' of everyday conversation may not boast quite the same watertight logic as strictly mathematical ones, but the standards set by mathematical proofs can elevate our discourses. If flawed human reasoning was an evolutionary necessity, then the flawless nature of mathematical proofs can counter the broken arguments that pervade our social interactions. They can also train us to spot the fallacies of pattern-hungry algorithms.

A proof and a lie

To appreciate the uncompromising rigour of mathematical proofs we will look at one of the most well-trodden results of the subject. You know it as *Pythagoras's theorem*. It states that for any right-angled triangle, the lengths a, b and c are related by the formula $a^2 + b^2 = c^2$, where c is the length of the *hypotenuse* of the triangle (the diagonal) and a, b correspond to the lengths of the other sides (*adjacent* and *opposite*, if you care to name them). Among other things, the theorem gives us a mightily convenient method for calculating the distance between two points in terms of just the 'width' and 'height' that separates them.

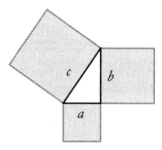

The most familiar example is the triangle whose sides have lengths 3, 4 and 5: you can verify that $3^2 + 4^2 = 5^2$. According to Pythagoras, $a^2 + b^2 = c^2$ always, definitely, absolutely holds whenever a, b and c correspond to the lengths of the sides of a right-angled triangle. The significance of Pythagoras's theorem is that it holds for all right-angled triangles, large or small, blue or pink – all infinitely many of them. To proclaim any statement for an infinite collection of objects is audacious. Mathematical reasoning emboldens us to make such claims because it pierces through finite limits. It does not rely on patterns or probabilities – only on the surest of logical leaps. Our reward is eternal truth; faultless logic cannot be undermined by any experiment, now or in the future.

Instances of the formula were uncovered long before the Greeks entered the fray, and it probably wasn't Pythagoras himself who issued the first known proof.[35] One argument proceeds by arranging four copies of any given right-angled triangle in two different ways, forming two squares. Note that we're not restricting ourselves to any particular right-angled triangle; the steps that follow will apply to every single one of them.

The key idea is that both of the large 'container' squares have the same overall area (because their sides are of the same length), and both contain four copies of our triangle. The only difference between these large squares is in how the four

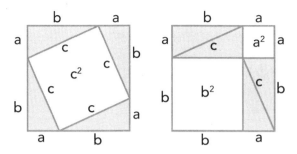

triangles within them are arranged. This means that the white component of each large square must take up the same space (the same area). The white space in the left-hand square is itself a smaller square with area c^2. As for the right-hand square, the white space comprises two yet smaller squares of areas a^2 and b^2 respectively. Since the white spaces account for the same area, it must follow that $a^2 + b^2 = c^2$, which is exactly what we set out to demonstrate.

The proof illustrates the nature of mathematical discovery. The result itself didn't just pop into being; it was discovered (several times over) through playful exploration. The same is true of the proof – the clever rearrangement of triangles only seems obvious in retrospect. Mathematicians do not work their way through proofs at a steady clip, smoothly dispatching one proposition at a time. Rather, they doodle and dawdle, often venturing down blind alleys, almost always erring and self-correcting before the key insight finally dawns on them.[36] Mathematical proofs usually contain a central idea that isn't immediately obvious but that emerges through considered study (or even just play). The core principle driving this proof is *invariance*: by keeping one thing fixed (the area of the large square), we can create equivalences between the remaining parts. It's a strategy we can take forward to other proofs, above and beyond right-angled triangles.

It is significant that a simple visual representation, combined with the deft tactic of computing the same area in two ways, was all that was needed to take control of one of the most dreaded results in all of school mathematics. Visual proofs can illuminate truths without compromising on rigour.* Yet the above proof of Pythagoras's theorem is only one way of seeing its truth. Elisha Scott Loomis, a maths teacher from Ohio, spent a lifetime gathering no fewer than 371 proofs in a single volume.[37] That is 371 representations of the same underlying truth – some visual, others abstract, and every one a distinct lens through which to understand this particular geometric property of triangles.

Mathematicians are pluralistic when it comes to proofs, seizing upon a wide range of representations to make their case. The unifying force behind proofs is logic; beyond that, it is fair game to employ symbols, pictures and whatever other means to forge cast-iron truths. There is a takeaway here for non-mathematicians, too: to strengthen the foundation of our presumed truths by interrogating them from as many angles as possible.

Because mathematical proofs operate on strict logic, the mere sight of a contradiction brings all arguments to an abrupt halt. Even the tiniest omission of reasoning can lead our arguments astray, sometimes landing us on absurd conclusions. To

---

*One of my favourite mathematics books, *Proofs without Words*, is in fact a Japanese translation. The friend who gifted me the copy knew full well that Japanese is not in my repertoire, but the premise of the book, as the title suggests, is that it is possible to demonstrate mathematical truths without any dependency on a native language. The book lives up to its title: I could make no sense of the Japanese symbols, but this did not hold me back from absorbing the proofs, which rested on visual arguments and standard mathematical notation.

demonstrate just how absurd, I am going to attempt to *prove* to you that *2 = 1.*

1. I will start by noting something we all know to be true: $1 = 1$

2. Next I will subtract 1 from both sides of my equation, rendering:

$$1 - 1 = 1 - 1$$

3. I note that 1 is the same as 1×1 (which I'll write as 1^2) so I can turn my equation above into:

$$1^2 - 1^2 = 1 - 1$$

4. Now I'll do something that all high school students should be familiar with: the left side is the difference of two squares and is the same as $(1 - 1) \times (1 + 1)$, giving:

$$(1 - 1) \times (1 + 1) = 1 - 1$$

5. Now let's divide both sides of the equation by the term $1 - 1$:

$$\frac{(1 - 1) \times (1 + 1)}{(1 - 1)} = 1$$

6. And finally, we can cancel the 'like' term in the top and bottom of the fraction (something we learned at primary school), leaving:

$$1 + 1 = 1$$

7. Well, $1 + 1 = 2$ as we all know, so this must mean $2 = 1$.

This is what I set out to prove. At this stage, you have two choices. You can either accept, somewhat boldly, the conclusion that $2 = 1$, or you can attempt to uncover the flaw in my argument. We will retain some measure of sanity by adopting the latter position, and by critically asking *why?* – or, in this case, *why not:* where does the argument defy sound reasoning? This painstaking process of debugging and self-correction is a core part of the mathematician's enterprise.

The opening gambit is uncontroversial: all I did was claim that the number 1 is equal to itself. From there, I made a few standard manipulations, and you can easily verify the legitimacy of steps 1–4 (step 4 requires familiarity with algebra but is fair game). We arrive safely at

$$(1 - 1) \times (1 + 1) = 1 - 1$$

From here, it is just a couple of small leaps to the outrageous finale: dividing by $1 - 1$. This step also seems fine: another standard operation applied to both sides of an equation. But *seems fine* is a far cry from *definitely, absolutely, 100 per cent legitimate*. Look again at the term we divided by, $1 - 1$. This is just a convoluted way of writing 0, of course.

Dividing by 0 is the cardinal sin of mathematics, a forbidden fruit that we are mandated by the laws of arithmetic to avoid (this warrants another *why?*). It takes an alert mind to notice that we were dividing by 0 in the fifth and sixth steps. The devilry of this proof is that its syntax conceals the illegal manoeuvre. (I could be more devilish still by replacing the numbers with letters, but will spare you the song and dance.)

We knew that something was awry; the preposterous

suggestion that $2 = 1$ prompted our search to find the flaw in our reasoning. Proofs can break down for a number of reasons: faulty assumptions, incorrect inferences and fuzzy handwaving among them. Interrogating proofs sharpens our capacity to detect the subtlest argumentative flaws.

With all that mathematical proof has to offer as a conditioning tool for the mind, we may well ponder what roles computers have to play in this enterprise. Proof has been of interest to the computing industry for decades.[38] It is a means of verifying that software programs can be trusted to behave as intended: that they won't break down or set off explosions when we need them the most. But now computers are eating their way into mathematical proof, forcing us to examine our notions of trust and certainty.

In what ways can computers aid or inhibit our search for permanent truths, and to what extent might they create proofs of their own? And have we been too quick to dismiss their potential as reasoning agents?

Proof by computer

Mathematical truths are compiled in three stages: a conjecture is made, it is proved or refuted, and the resulting argument is then verified either way (with several iterations as we deal with failed attempts, false beliefs and new insights). Computers are sinking their teeth into each of those stages.

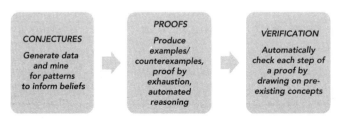

CONJECTURES
Generate data and mine for patterns to inform beliefs

PROOFS
Produce examples/counterexamples, proof by exhaustion, automated reasoning

VERIFICATION
Automatically check each step of a proof by drawing on pre-existing concepts

To start with, computers can serve up clues in the quest for new truths by generating numerous examples from which we can develop our intuitions and test our hypotheses. Suppose you want to know how likely it is that you would flip exactly five heads in ten throws of a fair coin. A novice programmer can create a simulation that performs ten flips a million times over. You will find that you get five heads in around 250,000 of those simulations (probably not exactly, but uncannily close). You therefore develop a hypothesis that the probability of landing five heads is a quarter. You still have to find a robust proof that does not rely solely on experimental data, but having a plausible hypothesis at hand is a winning position to start from. Computers can go further still by analysing data and detecting patterns that may not be obvious to humans. As mentioned in the introduction, the tools of machine learning are being aimed at problems across a range of mathematical fields. Mathematicians can refine their conjectures, and even develop new ones, on the back of new evidence brought forward by machines.

We should be careful not to rely on patterns alone as a source of evidence. It doesn't really matter how powerful computers become: their calculations remain limited by the physical bounds of our universe, which means they will never account for the complete set of truths of *really large* numbers through sheer computation alone. This is no small limitation because really large numbers take on a life of their own, often defying truths we take for granted at the lower end of the number line.

A case in point is prime numbers, whose behaviours are about as predictable as British weather in spring. There are infinitely many primes, and what makes them so compelling is that there is no discernible pattern that determines exactly when the next one will occur. Primes are the building blocks of arithmetic. Every number can be expressed as a bunch of prime numbers multiplied together: $30 = 2 \times 3 \times 5$, $126 = 2 \times 3 \times 3 \times 7$ and $13,143,123 = 3 \times 3 \times 7 \times 7 \times 29,803$. What's more, a number's 'decomposition' into primes is always unique (up to the order of multiplication). In 1919, the Hungarian mathematician George Polya pondered on the fact that some numbers are decomposed into an even number of primes while others are broken into an odd number: 30 has three primes in its decomposition and 126 has four (we count the 3 twice). Polya wondered, as mathematicians do, which occurred more often – even decompositions or odd ones? If you check the numbers from 1 to 10, you'll find that 6 are odd-type and 4 are even-type. If you go as far as 100 you'll find that 51 are odd-type and 49 even-type. Among the first 1,000 whole numbers, 507 are odd-type and 493 even-type.

A pattern had emerged, and it made Polya speculate, *conjecture*, that up to any threshold, there will always be at least as many odd-type numbers as even-type ones. Polya's conjecture was checked and verified up to a million; the evidence was stacking in his favour. Yet in 1962 another mathematician, Russell Sherman Lehman, found an example that contradicted Polya: he showed that up to the number 906,180,359 (that's over 900 million), the number of even types exceed the number of odd types (by just one, it turns out). Larger numbers tilted the balance in favour of the even types.

These days you can easily write a computer program that checks Polya's conjecture and informs you that it eventually breaks down at around the 900 million mark. But there are some conjectures that hold for dizzyingly high numbers:[39] so high, in fact, that if you transformed all the matter of the universe into paper and ink, you would still run out of writing materials before you reached the all-important failure points. In these instances you might be forgiven for not waiting until your computer program unearths the colossal counterexample (after all, it took just five iterations of the circle regions for us to subscribe to a false hypothesis). In the face of such overwhelming evidence, you might decide that a rigorous mathematical argument is not worth your while and succumb to a mistaken conclusion.

While today's computers are gaining power at an exponential rate, some numbers are just too large and unwieldy to handle. Computation cannot cut through the sheer size of these numbers in the way that mathematical reasoning can.

Computers are also making their mark in the construction of proofs, altering the way we establish permanent truths.[40] The proof of the four colour theorem, remember, came in two parts: the human contribution from Appel and Haken to reduce the problem down to around 2,000 map configurations, and the might of computers to work through those cases.[41] This *proof by exhaustion* method is a perfectly valid argumentative strategy. When Haken's son presented the proof to an audience, they split into two groups: the over-forties could not bring themselves to accept that a significant part of a proof could be done by a computer. The under-forties were equally sceptical of the 700 pages of handwritten arguments and calculations from the two human authors. Complicated proofs ask for a leap of faith from their readers; trust must be granted to an *other* to negotiate the final detail. Placing your faith in computers is surely no more objectionable than placing it in fellow humans, especially when the scope for error is vastly reduced.

Computers can also help *disprove* conjectures by generating counterexamples. The Swiss mathematician Leonhard Euler claimed in 1769 that the following equation has no positive whole number solutions:[42]

$$a^5 + b^5 + c^5 + d^5 = e^5$$

Euler is among the most prolific figures in mathematics folklore; it is said that one would have to devote one's entire adulthood just to pore through his manuscripts. Nevertheless, he took this particular conjecture to his grave, having failed to produce a proof. With a computer at hand, one can check through an enormous number of possibilities, which eventually leads to the following case:

$$27^5 + 84^5 + 110^5 + 133^5 = 144^5$$

This counterexample surfaced some 200 years after Euler made his conjecture. In addition to taking a few spins in his grave for his egregious error, Euler might lament the fact that he did not have the calculating tools to disprove his own idea. Today's mathematicians wield the computational means to lay siege to some of the most stubborn problems involving arithmetic.[43]

Computers are even being marshalled to construct proofs from start to finish. As AlphaGo stitched together plays that wowed expert human players and mathematicians alike, mathematical proofs became a natural target for the Go conquerors. The connection between formal mathematical proofs and intricate board games runs deep. To construct a proof, we need a set of starting assumptions ('axioms') and a list of rules for making logical deductions. If mathematics was a game, these ground truths would make up the starting position on the board and the rules of logic would correspond to the rules of the game. A proof in mathematics is, then, analogous to a series of legitimate moves in a board game.

Once you're comfortable with the board game analogy, it's

not a stretch to conceive of a computer program that knows the starting positions – the axioms – and is able to crunch through different combinations, using any accepted rule, to see what other, more complicated truths it might arrive at. The prospect is nothing short of the automation of mathematical discovery, and it's one that many mathematicians are paying serious attention to.[44] Different groups have banded together to hardcode mathematical concepts – definitions, simple results and even major theorems – in terms a computer can understand. As the database of formalised mathematics grows, so too does the scope for discovering new and novel truths. Humans may take the hot seat and interact with these theorem-proving systems, but it is an enterprise that is well suited to algorithms.

The Google DeepMind team, for instance, has applied its machine learning methods to the Mizar database of mathematical proofs. The approach is novel; these algorithms do not themselves reason, but they can reproduce perfectly reasoned arguments by imitating the structures of known proofs and extending them to establish new theorems. It is not even just a matter of rehashing new proofs to old theorems; in some cases these algorithms can establish whole new results.[45] There's a fusion here of classical AI (the formal proofs contained in these databases) and the contemporary paradigm of machine learning.

On the other side of mathematical discovery is *proof verification*. Once a mathematical argument has been presented, it needs to be meticulously checked; the status of eternal truth must be earned. The acceptance of mathematical proof has traditionally depended on the scrutiny of experts in the field – the so-called *peer review* process that underpins research publication. This is sometimes termed 'proof by authority' – but whose authority? For the greater part of human history, checking proofs has been a human endeavour. But just as we invented tools to offload the burden of large, tedious calculations, mathematicians are now turning to *proof verification systems*: computer programs that evaluate every step of a proof, using the rules and concepts that have already been specified to them. To go back to the board game analogy, this is much like taking a configuration of pieces mid-game and trying to trace it back to another configuration that is already known to arise from legitimate gameplay.

The mathematician Thomas Hales brought this technique into prominence when, in 2014, he announced a formal proof to a 400-year-old problem called the *Kepler conjecture*.[46] This problem states that the densest way to pack spheres is through 'face-centred cubic packing' (in simpler terms, the way you usually see oranges stacked in a crate). Street vendors have long suspected as much, and Hales finally delivered a proof – long, complicated, every step confirmed by a computer. Hales had actually delivered a proof almost two decades earlier, which used computers in a different way – to work through a large collection of specific cases (proof by exhaustion again, just like the four colour theorem). The original proof was published in the *Annals of Mathematics*, but the editors, incapable of checking every calculation, were only willing to declare the proof as '99 per cent certain'. Looking to bridge that gap, Hales took to developing a proof that left no doubt. In his 2014 upgrade, the computer was now checking through statements of logic rather than just rattling off specific instances of the theorem.

These 'computer-assisted proofs' are becoming increasingly common in the furthest frontiers of mathematics,[47] where research papers can span hundreds of pages and contain symbols and notation that humans themselves struggle to grasp.[48] Opting for a human checker is a philosophical stance rather than a practical one; if the goal is to maximise certainty, we must acknowledge that humans are more prone than machines to letting a subtle assumption slip through the net.

Automated mathematicians

This talk of automating proofs may come as unsettling news to mathematicians who might have assumed that proof-making was their calling card. Before reaching for the panic button, though, a cursory look through the repository of Mizar proofs suggests something is lacking in this style of argument. The validity of the proofs is perfectly sound, and their concision – typically just a few lines – is impressive. Yet the blunt syntax that is a staple for computers robs us humans of any kind of enlightenment. We can convince ourselves, through these short, symbol-laden proofs, that Pythagoras's theorem and other results hold. But gone are the deft manoeuvres, clever ploys or central insights that we can take to other areas of mathematics.

The mathematician and writer Marcus du Sautoy likens automated theorem provers to the library of Babel, Jorge Luis Borge's fictional trove of all possible 410-page books.[49] The failure to distinguish between standard (or even bog-standard) texts and truly creative ones offends our literary sensibilities. Likewise, mathematicians aren't to be found on the assembly line of theorem factories, mindlessly churning out one result after another. They seek out the essence in proofs: the core insight that reveals some deeper truth or connection to other ideas.[50] This may explain why I've yet to encounter a mathematician who feels threatened by the slings and arrows of automation. As Henri Poincaré put it a century ago: 'A machine can take hold of the bare fact, but the soul of the fact will always escape it.'[51]

Humans process their truths beyond the abstractions of symbols. In one study, subjects were asked to examine the syllogism 'All A are B. All B are C. Therefore, all A are C.' We've met this construct already as the *modus ponens* rule. The subjects were also asked to assess the validity of another statement: 'All dogs are pets. All pets are furry. Therefore, all dogs are furry.' It is *modus ponens* once again: identical in logical structure to the first statement. This time, of course, there is rather more tangible meaning to the words. We all have preconceptions of furry pets that we lean on to evaluate such statements. Using fMRI imaging, the researchers confirmed that we deploy different neural networks within the brain to assess the truth of each statement.[52] Unlike computers, we can process statements in terms of what we know of the world. And mathematicians, likewise, prove their theorems in the context of everything that came before, and everything that might follow. No proof is an island.

The representation of a proof matters just as much as the truth it establishes. The mathematician Paul Erdös spoke of the 'book', a divine collection of the most elegant, satisfying arguments.[53] A good proof should animate the essence of a particular truth. It should read as a compelling narrative. Just as the best stories delight us with surprising revelations, the most satisfying and memorable proofs are the ones that hook us to their twists and turns (stories, recall, also aid memory as a chunking mechanism). They unravel like an Agatha Christie thriller – watching the detective deftly piece together the clues is as gratifying as discovering the actual identity of the perpetrator. The infusion of logic, meanwhile, prevents the proof from drifting towards falsehoods. As we have seen, storytelling itself may lead us away from valid conclusions as a raft of narrative biases kick in. A mathematical proof protects against such deficits by adding logic to the mix as a kind of narrative arc – it is the gatekeeper of each touchstone of a mathematical argument.

This ability to blend logic with evocative representations doesn't always carry over to the 'real' world, of course. Mathematicians are hardly immune to logical fallacies when discussing politics, religion and other everyday topics that interact with our strongest beliefs and values. Distorted worldviews arise either when we adopt shaky foundational beliefs (more on this in the next chapter) or when we allow our biases to overwhelm our reasoning. But it bears repeating that proofs serve as a conditioning exercise for the mind. Just as time spent at a yoga retreat can awaken our spiritual senses (and increase our flexibility for good measure), time spent engaging with proofs strengthens our capacity to develop and refute arguments in both mathematical and non-mathematical realms.

The requirement that proofs appeal to our subjective notions of elegance renders mathematics an art form as much as a science, a view to which many mathematicians have subscribed. These qualities elevate proofs beyond cold logic, eluding the symbol-crunching ways of machines. Computer-generated proofs exemplify the 'ugly mathematics' that the mathematician G. H. Hardy disregarded – for him, *beauty* was the first test of mathematics. His criteria for beauty rested on the 'seriousness' of an idea, economy and a sense of unexpected inevitability. Others have elaborated on what might otherwise be dismissed as a nebulous appeal to aesthetics.[54] A recent study from Yale suggests that even non-mathematicians can intuit mathematical beauty just as much artistic beauty, and that there is consensus among them for what constitutes a beautiful piece of mathematics.[55] The shortcoming of the Mizar Project's method of proof is that it casts beauty aside:

there is no storytelling; no narratives, no characters or sub-plots that guide us on journeys full of surprise and delight. And while efforts are under way among AI researchers to render computer-generated proofs more human-like,[56] for now, those proofs offer little beyond bloodless threads of logic.

Logic and emotion entwined

Let us take stock of where the last two chapters have led us. All arguments exist in a social context where the goal is not only to convey ideas, but to justify them and persuade others of their truth. To achieve all that in one thrust, an argument must appeal to both logic and emotion: the first to secure the objective truth of the proposition, the second to express that truth in terms that shifts people's beliefs in the right direction. Mathematical reasoning is a rigorous mechanism by which we separate actual truths from apparent ones. It addresses the logical dimension of arguments. The strongest arguments – those that evoke and deliver wisdom and insight on top of mere truths – also rely on the rich tapestry of representations that the human mind is endowed with. They make use of symbols, pictures, stories, analogies and other illustrative tools to reveal the big ideas of an argument rather than its conclusion alone. The messy and complex nature of our world can only be made sense of through the fullest range of representations at our disposal.

AI is still grappling with truth. The pattern-matching methods of machine learning are highly opaque, have no contextual basis for their judgements, and are restricted to memorisation as the primary means of making inferences. Reasoning is not yet in their circuitry. For a computer to think and act logically, the 'classical' systems of AI will need to be included after all, with hard-coded rules of inference, penned by humans, bootstrapping the reasoning capabilities of machines.

Many AI systems today already adopt such hybrid approaches; autonomous vehicles do not need to run over a million humans before acknowledging their fault. They 'know' that this action is prohibited because of the rules laid down by their human programmers. The DeepMind approach to automating mathematical proofs is a particularly novel blend of old and new AI that rejects the *tabula rasa* approach of learning everything from scratch, and instead seeks to feed off preexisting, human-generated proofs. A similar approach is at the heart of Dr.Fill, an automated crossword solver that uses deep learning to read puzzle clues and serve up possible answers, which are then ranked using an 'old-fashioned' algorithm that looks for things like the length of each word and whether they create conflicts in the grid. The combination has proved potent enough to win the American Crossword Puzzle Tournament.[57] When AI systems fuse multiple facets of intelligence in this way, we may well witness the stunning successes of AlphaGo in other, more profound spheres like mathematics and science.

Even then, humans have much to cling on to. We can hardly lay absolute claim to reasoning when our own cognitive systems are replete with in-built biases. As we project our mental selves onto machines, we would do well to remember our own faults, to prevent those systems from deepening our subtlest prejudices. But the human mind has also endowed itself with methods of reasoning that incorporates both logic and emotion. Our arguments are as diverse as they are elegant. They do not just deliver hard truths but also wisdom and insight: traits that are not so easy to encode.

4

IMAGINATION

Why spoilsports deserve more credit, how
mathematics gets reinvented, and the
truths computers will never discover

The board game Monopoly is banned at my house. My wife issued the prohibition several years ago when she could no longer endure the arguments that erupted every time I sat down to play with her siblings. The discord invariably stemmed from differing interpretations of the rules. Trading properties is the lifeblood of the game, but on several occasions her family members have, when driven by desperation, teamed up midway through a game, one player handing the other their prized assets for lowly sums. The collusion is patently obvious (to me at least) but to hear my fellow players tell it, this type of 'cooperation' is all part of the game. In these moments, I can only appeal to the *spirit* of the game which, as its very name suggests, is predicated on individualistic competition. No collaboration, no collusion: just cut-throat decision making that serves one's own interests. Since there is nothing in the rules to explicitly forbid these conspiratorial tactics, though, play goes on. At some point, we spar over other invented rules, like the one that rewards players who land on Free Parking with all the fines and taxes that have accumulated on the board. I bristle at this innovation because it adds more volatility to a game that already privileges luck over skill. Other players disagree; they want to live a little, they say, and allow for the extra element of uncertainty. On more than one occasion we have had to declare a contest void, recognising that we were essentially playing different versions of the game. A failure to agree on the rules renders competitive gameplay impossible – hence the ban.

Monopoly epitomises the ways in which small variations in our ground truths can result in wildly different interpretations of a situation.* This isn't just true of games. Much of the diversity of human thought can be traced to differences in our core beliefs and values – the everyday 'rules' we each subscribe to. Two people can disagree and both be logical: the strength of each person's argument may be equal, but they have simply proceeded with different beliefs or assumptions.[1] The psychologist Jonathan Haidt has developed a theory of 'Moral Foundations' based on six core belief systems that he believes can explain our political and religious divides.[2] These include care/harm, loyalty/betrayal, fairness/cheating, authority/subversion, sanctity/degradation and liberty/oppression. We can better empathise with opposing viewpoints when we understand how they derive from differing interpretations of these beliefs. In the United States, for instance, whereas the left tends to ground its understanding of *fairness* in notions of equality (everyone deserves prosperous outcomes), the right views the same concept in terms of proportionality (you get out what you put in). From those axiomatic definitions, it's easy to see why liberals tend to favour social welfare programmes and higher

---

*In fact, the original version of the game was designed to expose the injustices of concentrated property ownership. It was called 'The Landlord's Game' and included elements such as a land value tax. Only later did the 'monopolist' ruleset prevail, in which property owners are rewarded with rent payments, and bankrupting opponents is the express aim.

taxes, while conservatives tend to advocate small government and deregulation. Each set of views is a natural, even logical, consequence of core belief structures.

Despite the widespread tendency to declare our beliefs as stone-cold proclamations, immune to counterarguments, there is something liberating about stepping back and tinkering with our assumptions.

In the previous chapter, we compared humans and machines based on the ways they each combine a given set of truths to form new ones, arguing that humans bring a distinctly aesthetic quality to their arguments. It's one thing to compare two chefs based on what they can cook up from a given set of ingredients, however, and quite another to give them licence to dream up new recipes by changing up the ingredients they are using. This chapter sets a higher bar for intelligence by emphasising the ability to *break out* of a given set of rules.

Computers are found wanting here. Even where they appear to display high levels of creativity, their outputs are bound by how you, the programmer, have specified the parameters within which they are allowed to operate. Computers necessarily operate within whatever constraints are given to them. Humans, in contrast, are attracted to possibilities that arise from breaking free of any such restrictions. Our rebellious instincts are sometimes our most creative; they are what gives rise to completely new ways of seeing the world, or new worlds, rather than derivatives of existing ones.

Mathematics, in this sense, qualifies as among the most creative of endeavours because it encourages us to flout the rules. The phrase 'thinking outside the box' has its roots in a popular maths puzzle (first introduced by recreational mathematician Sam Loyd in a 1914 compendium of puzzles)[3] that asks: without taking your pencil off the paper/screen, can you draw four straight lines that go through the middle of all of the dots below?

You may be familiar with the solution, which is found only by straying from the confines of the grid. The puzzle's signature quality is that it forces us to reject standard approaches.[4]

Whole branches of mathematics are conceived in much the same way. Every mathematical system is built on a set of ground truths that we accept as self-evident – its *axioms*. We saw in the previous chapter that every proof is constructed as a meticulous chain of logical arguments, which can be traced back to those foundational statements. A mathematical system is shaped by its axioms, much like the character of a person can be traced to their core beliefs. And just as a person's character can be reshaped as they change their beliefs and values, so too can we disrupt the mathematics we so often take for granted. Mathematics is an invitation to construct our own mental worlds by setting rules, only to shake up those worlds by subsequently breaking them.

What if?

Humans have long been dreaming up concepts and creatures that exist outside our physical reality. The oldest known figurative art object, discovered in a cave in the Lone Valley of south-western Germany, is the Lion Man of Hohlenstein-Stadel, a chimeric figurine that is half-human, half-lion. Sculpted around 40,000 years ago for purposes unknown, the Lion Man is a product of pure human imagination. It signalled a new cognitive ability for humans: *counterfactual reasoning*. We were not merely seeking to understand our lived everyday experiences, but also contemplating what other realities might exist. We were daring to ask *what if?* – an apparently simple question, and an antecedent to creative thought. When we're faced with a situation, the representations we choose in order to interpret that situation are surrounded by what cognitive scientist Douglas Hofstadter calls an 'implicit counterfactual sphere' – an array of variations that each diverge, even if slightly, from our perceived knowledge of the world.[5]

Throughout history, creative expression has been marked by an ability to break with convention. Hofstadter calls this 'jumping out of the system' or, to use his quaint shorthand, *jootsing*.[6] The most innovative artists are the most disruptive: those brave enough to venture beyond accepted precedent and to open their work to a greater sense of possibility. Through his nine symphonies, Beethoven uprooted the 'classical' rationality of Western music tradition and injected it with emotional impact. Caravaggio transformed Italian painting through the intense realism of his work, and a use of light and shade – chiaroscuro – that foreshadowed photography. James Joyce's *Ulysses* revolutionised the novel by introducing a striking plethora of styles, points of view and sub-literary genres into its framework. These artistic shifts are so contrarian that they are usually met with derision or confusion in the first instance, only to be accepted as perfectly commonplace later. New rules displace old ones, birthing new genres in the process.

Modern forms of entertainment, driven by technological

advances, are a licence to bring our wildest imaginings to life. My favourite movies are a roll-call of counterfactual thought (and echoes of dystopian fiction dating back at least as far as Mary Shelley's *Frankenstein*): what if machines became sentient and had objectives that threatened humans (*Terminator*, *The Matrix*)? What if a subset of humans developed mutations that gave them superpowers (*X-Men*)? What if time travel were possible (*Back to the Future*)? You may well question my taste in films, but you will hopefully appreciate the counterfactual mindset that is needed to create these worlds. Similarly, video games are now a portal to some of our most mysterious worlds. Game designers have the space to create a set of rules and explore the realities that follow those choices. These worlds often defy the rules of our ordinary lives – worlds where the laws of gravity and motion are upended as a deliberate design choice.[7]

In science, too, we rely on breaks with accepted rules to make great leaps of progress – what philosopher Thomas Kuhn famously called 'paradigm shifts'.[8] Disruptions to scientific thinking, and to artistic expression, require more than incremental advances. Mathematics is the most disruptive of sciences, one of the oldest means we have of creating new worlds, however outlandish. It puts us firmly in the director's seat, setting the stage with whatever axioms we see fit. Our prize is newly conceived worlds that are often at odds with the mathematics we have studied and the rules we have been taught to accept. The driving force of mathematical intelligence resides in this freedom to create – and recreate – entire areas of thought.

Euclid's *Elements*, we've seen, was the foundation of rigorous mathematical proof. All the geometrical statements in the *Elements* stem from Euclid's initial store of 'postulates'. The fifth of those postulates states, in essence, that if you take any

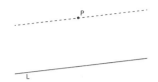

Euclid's fifth postulate – the dotted line is the
unique line parallel to the line L.

line, and any point off that line, then there is exactly one other line parallel with the original that passes through the point. That seems self-evident enough once you get past the convoluted wording, and Euclid even believed that it didn't need to be stated as a postulate because he would be able to derive it from the others. This statement, it can be shown, is equivalent to declaring that angles in a triangle add up to 180 degrees, or that parallel lines do not meet (claims that seem beyond any reasonable dispute).

As it happened, Euclid was never quite able to produce a proof of the fifth postulate, so was forced to call on it as an 'axiom of last resort'. In the centuries that followed, mathematicians tried to nip the fifth postulate in the bud by demonstrating that it is a logical consequence of the others, as Euclid had suspected. But the fifth postulate resisted these attempts, which eventually led a group of nineteenth-century mathematicians to contemplate the most daring of *what ifs* – namely: what if, after all, we allowed more than one line through that point to be parallel to the original line? What if we allow the angles of a triangle to total something other than 180 degrees?

Visualising these possibilities is a struggle for most people, but only because we are naturally drawn to flat surfaces, where Euclid's ideas hold up neatly. But what if we think instead about curved surfaces, such as a sphere, which is (roughly speaking) the shape of the Earth? If we draw parallel lines at right angles

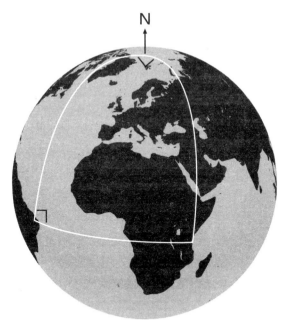

This triangle has three right angles, which sum to 90° + 90° + 90° = 270°.

to the Earth's equator, heading north (lines of longitude, for example), they would meet at the North Pole despite being parallel at the equator. If you go on to form a triangle between two points on the equator and the North Pole, then two of the angles are right angles, which means the sum of the three angles is in excess of 180 degrees. We now have a distinctly *non-Euclidean* geometry, where parallel lines do meet after all, and where triangles do not exhibit their usual behaviours.

This particular brand of geometry, known as *elliptic geometry*, should actually feel more comprehensible because it reflects the situation with a known physical structure, the Earth (understood by everyone bar resolute flat-earthers). Just as our understanding of the world is enriched when we reject the notion that it is flat, so mathematics can empower us with richer geometries that transcend the properties of a flat plane.

Creative thinkers must be prepared to suspend preconceived notions and physical norms. Why, after all, stop at the Earth's surface? The very first non-Euclidean geometry to be dreamt up was *hyperbolic*, situated in an even stranger world which tears up the fifth postulate by allowing infinitely many parallel lines to pass through the same point. Hyperbolic geometry is even more abstract, even more fanciful, but every bit as justified on its own terms. It pops up in models of special relativity as a way of describing the relationship between space and time, and it has also been posited as a way of describing social networks – 'useful' applications abound for those who seek them.

None of the alternatives to Euclid's fifth postulate renders the *Elements* false; they simply add to our worldview with equally valid frameworks for studying geometry, allowing for playful exploration with different building blocks. Janos Bolyai, a Hungarian mathematician who was among the first to dream up hyperbolic geometry, proclaimed in a letter to his father that 'out of nothing, I have created a strange new universe.'[9]

Can today's computers claim as much? There are now programs that can create photorealistic images of people who do not exist. They work using a clever technique called *generative adversarial networks*, which makes use of two models. The first, called the *generator*, is trained to create new examples that share a likeness with actual people. The second model, the *discriminator*, then looks at all the images together and tries to work out which are real and which are fake. It's a game of cat-and-mouse, where the first model is trying to deceive the second into believing its creations are real (hence the term *adversarial*). This process repeats until the discriminator is fooled about half of the time. At that point, the generator is producing fake images that are indistinguishable from real ones. This technology (which is also at the heart of deep fakes) is poised to transform photo-editing, special effects and industrial design. You will almost certainly have already been exposed to multimedia content that dazzles and inspires you, without realising that it is the handiwork of computers.

While a computer may dream up scores of new human-like figures after being fed images of real people, however, it could not conceive, say, the elves, dwarves or wizards that inhabit J. R. R. Tolkien's Middle Earth. Generative adversarial networks may help to fine-tune our fantastical imaginings on the big screen, but, while there are some efforts underway to direct them towards creating new worlds from old (replacing horses with zebras in videos, for instance),[10] those worlds are predominantly envisioned in human minds, which readily cast aside the familiar and ordinary.

Breaking the number system over and over

Mathematical truths are not cast in stone so much as they are cast in axioms. When we change the axioms, mathematical reasoning may guide us to truths that are unexpected and unintuitive at first, but unequivocally logical and powerful in their own right.

The number system that we're all familiar with (some tribespeople aside) is actually the result of several iterations of breaking through conceptual barriers. We know from Chapter 1 that our innate concept of exact number expires at around four – nature's design only revealed a small handful of exact, whole quantities to us. All else, starting with the whole numbers from four onwards, we invent for ourselves. To get from whole numbers to the fully fledged number system of today, humans have had to leap beyond everything we thought number could be and assimilate new extensions that were previously deemed unfathomable. The progressive evolution of the number system owes much to our willingness to depart over and over again from its accepted conventions.

Many of the struggles with numbers that students encounter in school arise because they are expected to draw on a concept or technique that disrupts their prior ideas of how numbers 'ought to' behave. Fractions are a classic example. When learning about whole numbers, we understand that they appear in sequence on a number line: 1 is less than 2, 2 is less than 3, and so on.

Fractions invert this ordering; our heads spin a little when we are first taught that ½ is larger than ⅓. Placing our whole numbers at the bottom of the fraction literally flips their ordering. Our intuitions are dealt another blow when we multiply by fractions. With whole numbers (greater than 1), multiplication always results in a larger amount – it amplifies. Multiplying by 2 doubles a number, multiplying by 3 triples it, and so on. Not so with their fractional counterparts: to multiply by ½ is to divide by 2, and thus to reduce a number's size. As we probe finer-grained classes of numbers, we have to accommodate new and often unexpected behaviours into our conceptual models.

Drowning in irrationals

No group of people has attached itself as fiercely to numbers as the cult of Pythagoras (he of right-angled-triangle fame). The cult lived by all manner of rules and regulations, which governed everything from their dietary regulations to their bedtime routines and how they put on their sandals. They were also proud disciples of numbers. Mathematicians and philosophers in equal part, the cult declared that whole numbers underpinned the very fabric of the universe. All things came from number.[11] In this arithmetical cosmology, fractions were permissible because they were easily described as one whole number divided into another, and easily constructed. But if a number could not be expressed as a fraction, well, that was far too unwieldy (for one thing, mathematicians would later show that any such number possesses a decimal expansion that runs on forever without repeating). This was tantamount to heresy as far as the Pythagoreans were concerned. The notion of these *irrational* monstrosities brought chaos to their view of an orderly universe.

According to legend, one of the cult's own disciples, Hippasus, stumbled upon the existence of an irrational number. He did so by playing around with Pythagoras's famed theorem, no less. Take a right-angled triangle whose shorter lengths are both 1. From the theorem, the length of the hypotenuse is the square root of 2. It is this number, so easily constructed, that Hippasus claimed could not be expressed as a fraction. It would be the last discovery Hippasus ever made; the cult apparently drowned him in an attempt to bury his treacherous discovery. Mathematical secrets don't stay buried for long, however, and another Greek, Euclid, would later include a proof of Hippasus's claim in Book X of his *Elements*.*

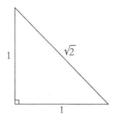

---

*Euclid adopts a proof by contradiction, which is a *what if?* type argument in itself. He starts by asking 'what if $\sqrt{2}$ exists as a fraction?' and then goes on to find a contradiction, thus disbarring any such fraction.

A host of other irrational numbers would surface through the centuries, and one stunning revelation is that these exotic beasts outnumber their rational counterparts. If you were to randomly drop a pin on the number line, you would almost certainly land on an irrational number rather than a fraction – that is, a number so strange that we can scarcely imagine it in the first place. The irrationals dispense with the notion that numbers are neat and orderly, forcing us to admit the existence of creatures that proved too wild for the Pythagoreans.

Zero: something from nothing

Legend has it that when Alexander the Great visited India, he encountered a wise man meditating naked on a rock. As the old man gazed upwards at the sky, the world conqueror asked him, 'What are you doing?' to which he replied, 'I'm experiencing nothingness. What are you doing?' Alexander said, 'I am conquering the world.' At this point they both laughed, each one considering the other a fool for wasting his life.

The concept of nothingness has received a mixed reception from civilisations in the past. Its mathematical representative, *zero*, was not inducted into the number system as a matter of course. Even the Greeks, despite all their mathematical sophistication, lacked any symbol for zero.

Ancient cultures each independently developed various symbols to represent the idea of zero. Babylonian cuneiform script included a double-wedge symbol for nothingness, while the Mayans denoted absence as a shell in their famed calendar system. Records of zero (denoted by a dot that later morphed into the 0 symbol we now adopt) also surfaced in third-century Sanskrit training manuals written for Buddhist monks.[12] In all these contexts, zero was not a number but a placeholder that denoted the absence of items. But zero as a *number*, that you could add, subtract and multiply with – an object in its own right rather than a placeholder – took more getting used to. The Mayans and Babylonians, despite using zero as a place-holder, had no notion of zero as a numerical *object*.

In India, practices such as yoga encouraged the emptying of the mind through meditation, and both Buddhism and Hindu-ism actively welcomed nothingness as part of their doctrines. This provided fertile ground for the concept of zero to grow in. The mathematician and astronomer Brahmagupta first described it as a number in his text *Brāhmasphuṭasiddhānta*, written in 628 CE. It would take another 300 years for zero to be accepted as a number in Europe, where cultural leanings did not sway towards notions of emptiness. In the early days of Christian Europe, religious leaders banned the use of zero on the grounds that, since God is in everything, any symbol that represents nothing must be the work of the devil.

Today, we can scarcely imagine a world *without* the number zero. It underpins everything from our notions of neutrality to the deepest cosmological questions of the universe's origins. When we encounter zero at school it brings little consternation – the doubt and suspicion of earlier generations has subsided.

Imaginary numbers

The numbers discussed so far, while troubling when first conceived of, can at least be understood in concrete terms. The square root of 2 has an unwieldy decimal representation, but it is simple to construct. Zero involves taking on board the concept of nothingness, but it is neatly situated at the centre of the number line, separating the positive numbers from their negative counterparts (while negative numbers themselves also require a conceptual leap). The next class of numbers we will examine is so daring in its scope, so removed from tangible notions of quantity, that they earn the label of *imaginary numbers*.

At school we are taught that square roots only exist for positive numbers. It makes perfect sense to think of 5 and −5 as *square roots* of 25 because both of these numbers, when multiplied by themselves, return 25. It is less clear what the square root of −25 should be: there is no obvious candidate because the numbers we are familiar with, when squared, always return a non-negative answer. Another, slightly more technical way of saying this is that while equations such as $x^2 = 25$ can be solved (this equation just says that there is a number x which, when squared, gives 25; we know this to be true of 5 and −5), equations of the form $x^2 = -25$ appear *insoluble*. This passed as conventional wisdom in mathematics for a long time; it simply did not make sense to square-root the negatives. Solving these types of equations was *not allowed*. Until it was, that is.

As far back as the ancient Greeks, mathematicians had flirted with the idea that such otherworldly numbers might exist – they just couldn't fit them into their existing conceptual frameworks. These numbers could not be grasped in the same material sense as, say, whole numbers, or even fractions. They appear to be mental concoctions. It was René Descartes who coined the term *imaginary number* as a derogatory descriptor; such was his discomfort with these entities. The Italian physician (and equal parts trader, gambler and astrologer) Gerolamo Cardano gave nourishment to the idea that negative numbers may have square roots. In his 1545 treatise *Ars Magna*, Cardano had noticed that in order to solve certain classes of equations, one had to break with usual methods and take the square root of negative numbers along the way. Cardano was puzzled by the sudden imposition of this new number type, which made a fleeting presence in some of his solutions. Not too long after this in 1572, in his work *L'Algebra*, the Italian mathematician Rafael Bombelli transformed the numerical landscape by showing that square roots of negative numbers can indeed be made sense of, so long as we are prepared to extend our conception of numbers.

The square root of −1 could now be willed into existence and given a name: *i* for imaginary, say. This number does not sit anywhere on the standard number line. But why settle for a line of numbers when you can have a whole plane? If real numbers are represented horizontally, we may as well display imaginary numbers vertically. Together, they form a two-dimensional plane.

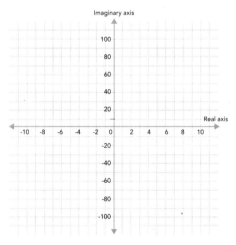

This is the *Argand diagram*, which displays all real and imaginary numbers. The number *i* is a unit up from the origin, just as the number 1 is a unit to the right. Similarly, the number 10*i* is ten units up, and −7*i* is seven units down. The Argand diagram is just a special type of coordinate plane, so if we now take any point in the plane, it has a horizontal component, *a*, and a vertical component, *b*. We write this number as *a* + *bi* and think of *a* as the real part of the number and *b* as the imaginary part. For example, we get the number 5 + 3*i* by going 5 units in the positive *real* direction and 3 units in the positive *imaginary* direction. This is called a *complex number*.

As you will have learned if you have taken mathematics far enough at school, we can perform standard operations on complex numbers: they can be added, subtracted, multiplied and divided. For example, to add the complex numbers 5 + 3*i* and 2 + 6*i*, we just add the *real* and *imaginary* parts, respectively, to get (5 + 2) + (3 + 6)*i*, or 7 + 9*i*. Most of the things you could ever hope to do to real numbers extend to their two-dimensional counterparts and, crucially, the complex numbers do not bring about any contradiction that would prohibit their use. In fact, with complex numbers in tow we can do much more, mathematically speaking. For one thing, we can laugh away the suggestion that equations like $x^2 = -25$ cannot be solved.*

Bombelli's imaginings caught the spirit of the Renaissance, a period that saw 'pure' mathematics – the more rigorous, abstract brand of the subject that is pursued without explicit regard for real-world application – soar, as mathematicians embraced their subject as a playground of ideas to be explored. Practical application was no longer the supreme motive for mathematical inquiry; the only entry criterion was a willingness to rigorously examine the consequences of whatever inventions the mathematician's mind could conceive. And yet, although imaginary numbers were dreamt up in the human mind and

---

*The fundamental theorem of algebra says that any polynomial – which is an equation involving powers of *x* – has as many solutions as the value of the highest power.

threatened to mark a complete break from reality (after all, there is no object in the physical world that obviously represents the number i), they have since proved ideal for describing a vast range of phenomena involving two-dimensional change, such as waves, electric currents and quantum mechanical equations. It takes an excursion into the mathematical world, which permits the existence of these strange but logical objects, to uncover some of the most powerful real-world representations.

Our ancestors could probably never have imagined the far-reaching conceptions of number that have developed over time. Every extension of the number system met with confusion, scepticism and even resistance in its day. But mathematics does not owe us comfortable truths. Slowly but surely, humans have integrated new ways of thinking about numbers. We have confronted anomalies outside our conventional wisdom and humbly recognised that the rules and regulations we accept as given may not tell the whole story. Our willingness to entertain new numerical possibilities rewards us with a wider lens through which to view the world.

The nineteenth-century mathematician Leopold Kronecker famously quipped: 'God made the integers, all the rest is the work of man.'[13] At the same time as betraying his own reluctance to accept thorny numerical constructs, Kronecker was (perhaps inadvertently) paying tribute to human ingenuity. We are willing to play at the boundary between what is and isn't considered possible, and we're equally willing to extend the presumed limits of our mathematical horizons.

One wonders if computers will earn credit in such adages: 'all the rest is the work of man, and beyond that the work of computers' perhaps. But a computer can scarcely grasp the subtleties of the number system that humans have slowly uncovered. Computers have no way of dealing with irrational numbers, for instance, save for executing known formulae for approximating them. Many irrational numbers are expressed as an infinite sum of terms, and a computer can get closer to the precise value by computing more and more of those terms, but this act alone doesn't imbue the computer with an understanding of the concept of irrationality. It's a huge leap to think that a computer could lead us to classes of numbers not yet conceived. That inventive power is not within the scope of rule-adhering machines.

The incompleteness of mathematics

The mischievous playfulness of the human mind is best exemplified through our attraction to paradoxes. Philosopher Willard Van Orman Quine defined a paradox as any 'conclusion that at first sounds absurd but that has an argument to sustain it'.[14] One of the oldest and best known paradoxes sprang from the mind of the Greek philosopher Zeno, who belonged to the Eleatic school, which considered change and movement to be illusory. In one telling of Zeno's paradox, the Trojan war hero Achilles is racing against a tortoise. Achilles allows the tortoise a head start of a hundred metres. Achilles runs very fast at a fixed speed while the tortoise hobbles along slowly, also at a fixed speed. Achilles will soon have travelled a hundred metres, arriving at the tortoise's starting point. By then, the tortoise will have moved along a small amount, say ten metres. Achilles then needs a bit more time to travel the extra ten metres. When he gets there, the tortoise has edged along. It seems that whenever Achilles arrives at where the tortoise once was, the tortoise has maintained a lead. So Achilles never catches the tortoise. This conclusion is patently absurd because Achilles is travelling faster than the tortoise – he will surely catch it eventually.

The value of paradoxes is that they impel us to revisit our assumptions and to refine and even reject them in servitude to sound argumentation. To resolve a paradox, we must address what went wrong in the argument. Zeno's paradox is resolved by recognising that time cannot be broken into discrete chunks in the way he has described. Thomas Aquinas provided one way of averting the apparent contradiction (foreshadowing the centuries-later development of calculus, which gave mathematicians a formal way of handling infinitesimally small quantities): 'Instants are not parts of time, for time is not made up of instants any more than a magnitude is made of points, as we have already proved. Hence it does not follow that a thing is not in motion in a given time, just because it is not in motion in any instant of that time.'[15]

The early twentieth century was a golden age of paradoxes, especially among mathematicians. A paradox poses a threat to the logical foundations of mathematics, but a system grounded in logic should be impervious to such threats and find ways of accounting for every apparent contradiction. One paradox that proved particularly potent came from the mathematician, philosopher and political activist Bertrand Russell. It is commonly framed as the 'barber paradox' and goes like this:

> *In a town, the barber is the 'one who shaves*
> *all those, and those only, who do not shave*
> *themselves'. Does the barber shave himself?*

There are two possibilities: either the barber shaves himself, or he doesn't. If he does, then he must be among those 'who do not shave themselves'. And if he doesn't, then he is among those 'who do not shave themselves' and therefore must shave himself. Either way, we find ourselves in a situation where the barber both shaves himself and doesn't shave himself – an outright contradiction. The paradox arises through self-reference: the barber has been defined in a way that refers to the barber himself, which ties the situation up in knots that cannot be untangled.

Russell conjured up this paradox at a time when mathematicians were attempting to ground their subject with rigorous foundations – a movement known as *formalism*.[16] The formalist movement enjoyed support from the likes of German mathematician David Hilbert, who initiated what became known as *Hilbert's programme* to establish a firm axiomatic basis for mathematics. To formalists, the essence of mathematics is inviolable logic. At first blush this view is not far-fetched; think back to the painstaking rigour of mathematical proofs. Formalists believed, perhaps intuited, that if the foundations of mathematics were properly framed, the whole subject could be kept free of the tyranny of paradoxes and contradictions.

As early as the 1880s, the German logician Gottlob Frege had already made a serious attempt to build mathematics from the ground up using only logic, by conceptualising all mathematical objects as *sets*. A set is just any collection of objects. Consider the number 3. In Frege's scheme, *threeness* is a property common to all sets containing three objects. The set of colours on the US flag, the set of primary colours and the set of blind mice of the nursery rhyme all share this threeness property, and the 'number' 3 is itself the set that contains all of these three-item collections.

The holy grail was to frame all mathematical statements in terms of these fundamental, abstract objects. Russell had realised that Frege's definition was too loose; his original paradox was actually framed in terms of sets. Russell asks you to consider a set – call it R – whose members are all sets that do not contain themselves. The question is, does the set R contain itself? You will soon realise that this is the self-referential barber paradox in disguise, where each possibility leads you to contradiction – anathema to the formalist ideals of set theory. Frege's framework was allowing for sets that were simply too large for logical rules to hold up. When Frege learned of his error from Russell, he hastily appended a postscript to his book, *The Basic Laws of Arithmetic*, saying: 'Hardly anything more unfortunate can befall a scientific writer than to have one of the foundations of his edifice shaken after the work is finished. This was the position I was placed in by a letter of Mr Bertrand Russell, just when the printing of this volume was nearing its completion.'[17]

Paradoxes were poised to plunge formalism into crisis, but Russell himself was unfazed: his own paradox simply gave him, and others, pause to reflect on what should and shouldn't be called a *set*. In the example above, R was simply too 'large' to qualify as a set. By barring such possibilities, he eliminated the resulting contradictions.

The formalists marched on, refining their definitions and rules in the hope of finding an axiomatic system that would be both consistent (i.e. free of contradictions) and complete (i.e. accounting for all mathematical truths). Euclid's system for plane geometry meets both requirements, but the same had not yet been established for systems of arithmetic. If a mathematical system could achieve both consistency and completeness, it would be strictly tied to logic once and for all, rather than to, say, intuition or other holistic thought structures. Intuition, after all, led to the absurdities of the barber paradox, deemed unfit for the sport of mathematics. Along with his contemporary Alfred Whitehead, Russell laid out his own approach (based on a slightly different system called *type theory*) in excruciating detail. Their tome *Principia Mathematica* is renowned for its complex notation: it takes several hundred pages of unrelenting notation to establish the fact that one plus one equals two. T. S. Eliot was among those who praised the work for its clarity and exactness, suggesting it was 'perhaps a greater contribution to our language [English] than ... to mathematics.'[18] In any case, the tome was the slowest but surest progress to a bug-free, all-encompassing version of mathematics. Or so they thought.

In 1931, Russell's formalist vision was dealt a vicious blow from which it would never recover, courtesy of the Austrian logician Kurt Gödel. Gödel's work is guided by another self-referencing paradox, the sentence 'this statement is false.' In much the same vein as the barber paradox, this statement can be neither true or false. What Gödel showed was that in any system sophisticated enough to contain the rules of arithmetic, you can produce the following statement, similar to that paradox:

This formula is unprovable in the system.

On the one hand, this statement is true because if it were false, then it would be provable, which would render it true (you may want to re-read that last sentence a few times). Yet the very statement of this truth tells us that it cannot be proved. One potential way out of this quandary is to turn this unprovable statement into an axiom, so that its proof is automatic. The substance of Gödel's argument is that the axioms upon which your system is based do not matter: there will always be another true statement of the form *'this formula is unprovable in the system'* that remains unprovable. It is much like an impossible jigsaw puzzle, where no matter how you arrange the pieces you always end up with gaps.

Gödel had created a gap between truth and provability, a

gap that could only be avoided if your system excluded elementary arithmetic. That is a severe constraint. The rules of arithmetic – how to define numbers and perform operations on them – are the basis of mainstream mathematics. The formalist ideal of complete and consistent mathematics would be limited to 'rogue' systems that deviate from our usual understanding of numbers. It would be akin to searching for a grand unified theory of the universe, only to realise the theory holds for no more than a handful of obscure galaxies. Gödel had demonstrated that no system containing elementary arithmetic can be both consistent and complete because there will always be statements, like that above, that cannot be proved or disproved – there will always be *undecidables*. This is the exact opposite of what formalism strives for. The entire edifice of Russell and Whitehead's work was shattered in an instant.

Gödel was not done – he later showed that even if a system containing elementary arithmetic is consistent, its consistency cannot be proved within the same system. The mathematician André Weil summarised it best: 'God exists because mathematics is consistent, and the devil exists because we cannot prove the consistency.'[19]

What Gödel established, contrary to the expectations of his time, was that logically consistent systems are small and boring. Mathematics does not yield to a foundational system from which all of its truths can be derived. It is broken at its core because no consistent system can account for all true statements.*

---

*If Gödel's statement strikes you as somewhat contrived, then it is worth noting that many other undecidable propositions have been uncovered, some of which correspond to concrete mathematical ideas. One instance arises in the highly pertinent area of machine learning, where researchers have sought

What incompleteness means for intelligence

This chapter espouses the essentially human virtue of breaking the rules. We have seen how altering the axioms of Euclid's plane geometry gives rise to entirely new geometries that are legitimate and powerful in their own right. We have also traced some of the histories of the number system, showing how humans have had to break through conceptual barriers to accommodate new numerical objects and behaviours. And we have seen that attempts to formalise systems containing arithmetic are doomed to fail because no consistent system can deliver a proof of every truth within that system. Mathematics is far more than a deductive discipline; it cannot be reduced to a collection of truths waiting to be proved in turn. The truth turns out to be more complicated and, at times, unprovable.

In one respect, Gödel's incompleteness theorems vindicate the rule-breaking mantra of this chapter. Since no axiomatic system can guide you to all proofs, our only recourse is to playfully tinker with axioms and examine the consequences. In their book *Gödel's Proof*, Ernest Nagel and James Newman write of incompleteness as 'an occasion, not for dejection, but for a renewed appreciation of the powers of creative reason'.[20] Gödel's arguments amount to saying that mathematical thinking operates on more than just precise, cold logic: it also demands some level of intuition and inventiveness. Quite often, an idea will require the seeding of new axioms, the creation

---

ways of making predictions about large datasets by sampling a small subset of data points. The question of whether a small sample is sufficient to make such extrapolations turns out to be equivalent to finding a set whose 'size' is somewhere between that of the integers (infinite and countable) and the real numbers (infinite and not countable) – a problem that mathematicians have known for decades is undecidable.

of new systems. Is this the kind of subtle intellectual work we can expect of machines? Nagel and Newman are among those who have expressed scepticism, suggesting that computers are tied to 'a fixed set of directives' that condemn them to the same limitations of formal systems. Where humans can show flexible reasoning and break up their own 'directives', computers are unavoidably stifled by the rules imposed on them.

There is also a detail embedded in Gödel's argument that may tell us something about the boundary between how computers and humans think. Consider again the 'true but unprovable' statement that Gödel constructs for a logical system (containing the rules of arithmetic):

This formula is unprovable in the system.

Gödel has managed to derive a truth about the logical system that the system itself is unable to establish. In other words, there is no way of 'seeing' this truth about the system unless we escape the rules of that very same system. In a famous argument put forward by the philosopher J. R. Lucas[21] and later revived by renowned mathematician and physicist Roger Penrose,[22] it follows that human thought systems cannot be wholly *algorithmic* – that is, reducible to a set of rules. If they were, the argument goes, then the mind would constitute a logical system and, by Gödel's argument, we ourselves would be unable to see the truth of its corresponding 'Gödelian statement'. But this is not the case: evidently, using some form of meta-mathematical reasoning, we manage to climb out of any prescribed system and are in fact able to see the truth of that statement. There is some wisdom or insight guiding us to truths not attainable within purely algorithmic constructs. And so, conclude Lucas and Penrose, machine intelligence, premised purely on such constructs, will never fully emulate the kinds of thinking espoused by us humans.

The Lucas–Penrose argument is not without its critics.[23] One counterargument is that our minds are so complex that we may have no way of formulating the Gödelian sentence they give rise to, rendering us no better equipped than machines to see the truth of the sentence.[24] Other arguments take aim at the assumption made by Lucas and Penrose that the system of human thinking is logical (or consistent). At the very least, we should reserve some suspicion for such an optimistic claim in light of the biases we uncovered in the previous chapter. Worse yet, by virtue of Gödel's second incompleteness theorem, even if we are consistent thinkers, we would have no way of proving this ourselves. In short, it may well be that we are *inconsistent* machines to which Gödel's theorems simply do not apply.

In any case, replicating human intelligence would require designing machines in such a way that they are not tied to any particular set of rules or behaviours – that is, machines that can entertain contradictions. Douglas Hofstadter sees no problem with this: 'It is no harder to get a computer to print out scads of false calculations ("2 + 2 = 5; 0/0 = 43", etc.) than to print out theorems in a formal system. A subtler challenge would be to devise "a fixed set of directives" by which a computer might explore the world of mathematical ideas.'[25]

The key question, it seems, is whether humans can program a machine in such a way that the machine can break out of its own system. The bar for human intelligence has been set by those who have rejected prescribed modes of thinking. By itself, logical manipulation gives us no way of confronting our belief systems. It has no means of challenging anomalies.[26] Creativity comes from discontinuity, from contemplating paradoxes and poking holes into our accepted ways of thinking. Humans advance new ways of thinking when we combine our logical disposition with a disruptive mindset, seeking out contradictions and resolving them.

In the aftermath of Deep Blue's victory over Garry Kasparov, the human grandmaster called foul (Kasparov was not known for being graceful in his rare defeats),[27] suggesting that the IBM team had intervened during one of the games to throw him off the scent. The exchange is revealing: one human accusing another of exploiting his psychological vulnerabilities by interfering with gameplay, in contravention of the rules. The accusation was levelled squarely at the human engineers behind Deep Blue, not the computer itself. Deep Blue, for its part, was incapable of breaking the rules. The inventiveness of Deep Blue and its more modern machine learning-driven incarnates such as AlphaGo remains confined to a world of fixed rules and regulations. AlphaGo and its like can deliver spectacular combinations of the rules and constraints they are fed, but perhaps the real bar for imagination is the ability of machines to contort the rules of the games they so effortlessly master, to 'think outside the box' and perhaps, just occasionally, to cheat. The 'spoilsports' who run amok in Monopoly may deserve more credit than I have granted them hitherto.[28]

5

QUESTIONING

Why mathematics is like play, the questions
no computer can answer, and the simple
trait that makes every child smart

It's not enough to know the Ultimate Answer to life, the universe and everything (which, as fans of Douglas Adams will tell you, is 42).* The question matters a great deal more. And yet, as Pablo Picasso put it more than fifty years ago, referring to calculating machines: 'they are useless. They can only give you answers.'[1]

Even the most fervent AI enthusiast would have to admit that the latter half of Picasso's caustic remark remains true. Useless they are not, but whether tasked to play a game of Go, drive a vehicle, or diagnose medical ailments, a computer's scope of inquiry is specified by us humans. They pursue answers only to the questions we pose of them. For a machine to *define* its own goals, it would first need to become sentient. While we await the literal awakening of the machines, asking questions will remain the preserve of humans.

It is widely thought that to develop human-level AI, machines must first be programmed with child-like intelligence, and then learn, as children do, by interacting with their environment (as

---

*As revealed in various adaptations of *The Hitchhiker's Guide to the Galaxy*, which also goes on to explain that the Earth is a supercomputer created to find the Ultimate Question.

opposed to having exabytes of information pumped into them at 'birth'). Alan Turing surmised as much in his seminal paper on AI: 'Instead of trying to produce a programme to simulate the adult mind, why not rather try to produce one which simulates the child's? If this were then subjected to an appropriate course of education one would obtain the adult brain.'[2]

If the AI community takes this aspiration seriously, it must figure out how to get computers to ask questions. If you spend just a few minutes with a group of children, you will be left with no doubt that questioning is the most innate of human skills. Infants are natural explorers. Even before developing their motor skills, they absorb visual cues from their environment and form hypotheses about the world. As children develop language, their observations trigger all manner of questions. Harvard child psychologist Paul Harris has put a number to it: his research shows that children ask around 40,000 questions between the ages of two and five.[3]

Even as adults, we possess an insatiable thirst for knowledge; what psychologists term 'epistemic curiosity'.[4] We're constantly navigating the terrain between what we know and what we don't: a sweet spot of exploration that arouses our liveliest curiosities.

Information-seeking can be extrinsic – we seek the latest insights on stocks to maximise income, and we check the weather to determine whether we should carry an umbrella on our afternoon walk. Machines are incentivised by external rewards, too. Robots that have been programmed with reinforcement learning algorithms take actions to increase some numerically defined reward. They'll roam their environment and make calculated choices around where to go next, or what to do next, based on what will score them points.

But information-seeking can also be intrinsic.[5] Unlike

machines, humans are drawn to questions that we deem interesting in and of themselves. We're so moved by cause-and-effect mechanisms, for instance, whether they have practical import or not, that we cannot help but be drawn to inquiry. If you ask a child to stack some building blocks, where some of the blocks have extra weights hidden in them, they will freely experiment to establish an atypical centre of gravity that makes their construction balance (essentially creating a model seesaw). Chimpanzees, on the other hand, show no interest in accomplishing the same task, presumably because they do not come equipped with the reasoning skills that render such tasks interesting.[6]

This chapter looks at the kinds of questions that pique our interest on their own merits, irrespective of external rewards. We can think of them as curiosity inducers, the most straightforward of which is, according to the psychologist George Loewenstein, the 'posing of a question or presentation of a riddle or puzzle'.[7] On that basis, mathematics is a powerful study of what makes the human mind tick.

The recreational roots of mathematics

Puzzles have been a staple feature of human interaction for thousands of years. Every now and then, they explode into popularity, holding masses of people in their grip. The earliest discovered collection of puzzles is the Rhind papyrus (dated 1650 BCE), a five-metre-long scroll that captures the Egyptians' penchant for measurement. It includes wide-ranging contributions to arithmetic and geometry, the Egyptians' very own decimal counting system and a collection of problems that demonstrate an extraordinary flair for unit fractions.* Amid all

---

*Fractions of the form $1/n$: $\frac{1}{2}$, $\frac{1}{3}$, $\frac{1}{4}$ and so on.

of the practical intent of the papyrus, Problem 79 has a teasing flavour: 'There are seven houses; in each house there are seven cats; each cat kills seven mice; each mouse has eaten seven grains of barley; and each grain would have produced seven *hekats* (an old unit of measure equivalent to about 5 litres). What is the sum of all the enumerated things?'[8]

The puzzle has undergone several iterations since (among them the well-known St Ives riddle, although that particular example requires no arithmetic). This is a problem of combinations, and it is rooted in playfulness rather than an authentic real-world scenario. Questioning for its own sake.

In England, the late eighteenth and early nineteenth centuries saw puzzles become mainstream. The conditions for recreational problem solving were ripe: printing had become cheap, leading to the widespread distribution of magazines, while the industrial revolution was giving rise to a 'leisure class' of citizens who had more time to indulge their curiosities. It was during this period that Lewis Carroll – the better-known alias of Oxford mathematician Charles Dodgson – wrote *A Tangled Tale*, a collection of ten humorous stories (or 'Knots' as he called them), each anchored to a mathematical puzzle. It's every bit as delightful as *Alice's Adventures in Wonderland* (which, to the attentive mind, also happens to contain its fair share of mathematically inspired riddles). Recreational puzzling crossed the pond in the twentieth century, most notably with the legendary Martin Gardner, who ran the 'Mathematical Games' column in *Scientific American* between 1957 and 1982. Mathematicians and puzzlers of all varieties converge in their admiration for Gardner, such was the diversity of his playful creations. Perhaps the highest praise has come from linguist Noam Chomsky, who asserted that 'Martin Gardner's contribution to contemporary intellectual culture is unique – in its range, its insight, and understanding of hard questions that matter.'[9]

The most prominent instance of recreational problem solving in recent decades hails from Japan: the Japanese have asked a few hard questions of their own by popularising their distinct genre of grid-like puzzles. You almost certainly know or have even attempted one of these: *Sudoku*. Faced with a 3 × 3 grid of 3 × 3 boxes, puzzlers are tasked with completing the partially filled grid in such a way that the numbers 1–9 fill every row, column and box contained within the larger grid. Sudoku actually originated in the United States, but it found a willing audience among the readers of a Japanese puzzle magazine, *Puzzle Communication Nikoli*,* which boasts over 300 different puzzles, most of which are generated by its readers. All of the puzzles are inspired by the grid format. Writer and puzzler Alex Bellos attributes the wild popularity of these puzzles in Japan to the culture's fondness of miniaturism, minimalism, refinement and craftsmanship.[10] The rules are deceptively simple, while the solutions require reasoning that is both logical and elegant. It is typical for magazines to defer to computers to generate multiple instances of their content, but every *Nikoli* puzzle originates from hand-crafted human effort. As a puzzler, you can 'feel the hand of the author' as you explore pathways through the grid.[11]

To a mathematician, the experience of doing mathematics is more akin to that of a puzzle enthusiast working through those grids than it is to the formal, stuffy presentation of the subject shaping its public perceptions. Gardner defined recreational

---

*The magazine's creator, Maki Kaji, happened to be a horse-racing aficionado. When he first published his magazine in 1980, he decided to name it after the favourite in that year's Epsom Derby, Nikoli.

mathematics as the part of the subject that 'includes anything that has a spirit of play about it.'[12] Arguably, the definition ought to extend to all aspects of mathematics – you will be hard pressed to find a mathematician, even one whose aims are practical, or whose ideas are rigorous and expressed formally, who does not delight in problem solving.

Much of the mathematics we take for granted today, even its most serious forms, can be traced to recreational roots. A single problem, considered interesting on its own terms, can motivate vast new concepts and even whole branches of mathematics.

The bridges of Königsberg and graph theory

In the eighteenth century, residents of Königsberg, a small Prussian town on the banks of the Pregel river, mused over a question during their long Sunday walks.[13] The river divided the city into four regions, and seven bridges had been built over the river to connect the regions. The residents wondered whether it was possible to trace a route around the city in such a way that they visit each of the four regions by crossing each bridge exactly once. The exact motivation for this puzzle seems unknown; perhaps there was an element of pragmatism among tradespeople to chart the most efficient path. More likely, the puzzle appealed to their problem-solving instincts and grew in stature as a solution remained elusive. The residents were unable to find a path, and were equally unable to demonstrate that it was impossible to do so.

In the nearby city of St Petersburg, the puzzle caught the attention of the prolific mathematician Leonhard Euler. Euler initially dismissed the problem as trivial, writing to a colleague that it 'bears little relationship to mathematics'. But Euler could not shake off his curiosity, soon afterwards confessing that the

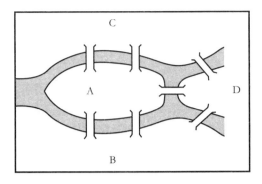

puzzle 'seemed to me worthy of attention in that [neither] geometry, nor algebra, nor even the art of counting was sufficient to solve it'. Euler, having exhausted the standard toolkit of mathematicians at the time, realised that solving the puzzle might motivate new mathematical concepts.

He could see that the puzzle resembled a geometry problem, but not geometry as it is usually understood, with measurements and calculations. The key idea behind this puzzle, Euler realised, was in the *position* of the four regions and the seven bridges that connected them. Euler created a new representation for the problem by thinking of the regions as points (or nodes) and the bridges as lines.

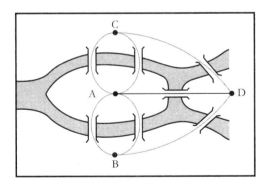

In Euler's mind, the puzzle amounted to drawing the following configuration of dots and lines without going over the same line twice, and without removing the pencil from the page:

You may have encountered puzzles of this type and found that some appear to be unsolvable. Euler's challenge was to show why this particular map could not be drawn in the prescribed way (which would then explain why the Königsberg round trip was impossible). Euler's idea was to consider the number of lines (bridges) that connect each point (region). Take any point that does not start or end the journey. If we are to make a journey, not removing our pen and not going over the same line twice, then every time a line goes into the point, another line must go out. This means that an *even* number of lines go through the point over the whole journey because adding two to itself always results in an even number. If we ever find a point with an *odd* number of connected lines, it must be situated at the start or end of the journey.

Now look at the Königsberg map again. Each of the four points has an odd number of lines connected to it. That's too many: clearly, at most, two points can be placed at the start and end of any journey. And that's why the Königsberg residents failed to come up with a legitimate route – it didn't exist.

By formulating and solving the puzzle in this way, Euler brought into being a new field of mathematics called 'graph theory'. Here, 'graph' means something different to the graphs you plotted in school: it denotes any arrangement of connected lines and points. Euler wasn't just solving a single problem for the residents of Königsberg, he was inventing whole new concepts and tools for studying connected groups of objects. Graphs turn out to be very interesting from a mathematical standpoint; once the impossibility of the Königsberg puzzle was established, Euler, and generations of mathematicians after him, set their minds to exploring other, similar, problems. Graphs crop up in all kinds of real-world situations as models of different types of networks. The tools of graph theory can be used to explore the structure of the human brain, the atomic arrangement of crystals, the hexagonal lattices woven by bees, the mega-wireless network of computers known as the internet, friendship groups on social media and the spread of infectious diseases. Graph theory has evolved into a vast field and an active area of research, yet it emanated from a simply stated puzzle that captured the imagination of humble residents going about their daily walk. It just took a mathematician to systematise the ideas that had piqued their interest.

A gambling problem and probability

A puzzle was also the trigger point for the development of probability, the study of chance events. Probability was not always accepted as a mainstay of mathematics; uncertainty was a thorny concept that, for those deferential to the very definite, predictable laws of nature, was not to be tampered with. The field was sparked in a famous 1654 letter from one French mathematician, Blaise Pascal, to another, Pierre de Fermat.[14] Pascal was a torn personality who flitted between gambling and religious practice. Somewhere in between, he found himself more than a little intrigued by the following thought experiment posed by a colleague:

Imagine that you and I are flipping a fair coin. We'll throw the coin five times and on each throw, you score a point if the coin lands Heads, I score if it's a Tails. Whoever scores more points across the five throws scoops the £10 prize money. The first three coins land Tails, Heads, Heads. At this point, the game ends abruptly (perhaps we lost the coin).

The question that bugged Pascal was: what is the fairest way to divide the prize money, given what we have observed? While the experiment is situated in the context of gambling, the scenario should strike you as fanciful: can you really imagine such a game being halted in this way? It has the flavour of a recreational puzzle and was compelling enough for Pascal to search for a solution that the mathematics of his time did not serve up. Several books had been published on gambling, but none had dared to make predictions of the future. That all changed in the series of letters exchanged between Pascal and Fermat. It was Fermat who figured out how to resolve the prize money – the solution may seem obvious to us now, given how accustomed we are to probability, but calculated speculation of future events was unheard of at the time.

Fermat reasoned that there are four possible future scenarios, depending on how the two coins land. For each of these possibilities, we can determine the hypothetical winner:

HH	–	4 Heads, 1 Tails	–	you win
HT	–	3 Heads, 2 Tails	–	you win
TH	–	3 Heads, 2 Tails	–	you win
TT	–	2 Heads, 3 Tails	–	I win

Since you triumph in three out of four cases, you should get three quarters of the prize money (£7.50), while I grudgingly accept the remaining quarter (£2.50).

With the puzzle solved, the field of probability was poised to go mainstream. Within a hundred years, mathematicians would invent mental tools to cope with all types of uncertainty. Johann Bernoulli applied these ideas to settle a range of legal disputes, such as how and when to divide up an estate when the owner is missing and presumed dead. His brother Jacob coined 'probability' as a term, formalising the calculations underpinning the likelihood of events. Other core tenets of the subject like the bell curve (used to model the distribution of several real-world characteristics such as height, weight, stock prices and test scores) and Bayes' Theorem (used to calculate the likelihood of one event based on the occurrence of other events) materialised soon after. A cottage industry of insurance providers entered the fold, profiting from the new enterprise of risk management. Fast-forward to the present day, and entire swathes of AI are underpinned by the same probabilistic ideas.

The mathematics that fuels so many real-world applications is rooted in thought experiments that might otherwise fill the pages of puzzle books. The usefulness of mathematics often emerges at a later stage, once concepts are firmly established, and owes a huge debt to the curiously minded mathematicians who get hooked on particular questions for the sake of puzzling alone, and who bring those concepts into existence.

Engaging with a maths puzzle or problem (the distinction barely registers with mathematicians – they delight in both just the same) can elucidate wider themes and principles that help to build up a picture of an underlying field, giving rise to more questions in the process. According to Pulitzer Prize-winning historian David Hackett Fischer, questions are nothing less than 'the engines of intellect – cerebral machines that convert curiosity into controlled inquiry.'[15] In mathematics, they have a generative effect: the solution of one problem begets many more. Mathematics is not as absolutist as many would claim: the resolution of a question isn't always a simple right or wrong answer. A good question is an expansive one: it motivates new definitions, concepts and theorems. Fields as wide and deep as graph theory and probability come into being when problems are linked together, bound by a common character.

So mathematicians are problem solvers on the one hand, puzzling their way through standalone curiosities, but also theory builders who seek out the trends and principles that govern the problems in their field.[16] Freeman Dyson, a mathematician and physicist, turns to the animal kingdom to describe these two approaches:

> Birds fly high in the air and survey broad vistas of mathematics out to the far horizon. They delight in concepts that unify our thinking and bring together diverse problems from different parts of the landscape. Frogs live in the mud below and see only the flowers that grow nearby. They delight in the details of particular objects, and they solve problems one at a time.[17]

The frogs-and-birds metaphor has echoes of philosopher Isaiah Berlin's classification of people as either narrowly focused hedgehogs (who 'know one big thing') and generalist foxes (who 'know many little things').[18] Dyson was a self-proclaimed frog, but was keen to emphasise that 'the world of mathematics is both broad and deep', implying a need for both types of creature: 'birds give it broad visions', he says, 'and frogs give it intricate details.'

Computers are frogs rather than birds. They'll fixate on

individual questions and answers without dwelling on how they relate to wider concepts. They may deliver spectacular and unexpected answers, but they lack the frame of reference to flag their results as spectacular or unexpected, or to tie together different solutions into unifying theories.

It's also not yet clear how computers might come one day to present the world with novel problems. Picasso was right to suggest that computers work strictly in the realm of problem *solving*, in servitude of the questions that we humans pose to them. We can put computers to the task of checking different configurations for the Königsberg problem, but it's difficult to imagine robots tottering about their daily excursions and thinking up such seemingly arbitrary problems in the first place, or feeling an urge to examine the apparent impossibility of a solution and originating new mathematics in the process. It is telling that the probabilistic foundations of modern AI applications emerged from the thought experiments of curious human minds. Until machines can somehow emulate the human drive for inquiry, it is far from obvious that they could ever produce mathematics for its own sake.

Where to point our 'telescopes and spaceships'[19]

Computers may not dream up problems, but they are helping us grapple with increasingly complex ones. Large datasets and superhuman computing power mean that we can model more of the world more quickly and reliably, from the minutiae of optimising a mobile phone package within budget constraints to large-scale, high-stakes climate forecasts. The human effort is in developing plausible models of these phenomena, before feeding them into our computers and then evaluating our assumptions based on what comes out.

This manoeuvre from the physical world to the computational one and back again is so far-reaching in scope that mathematician Conrad Wolfram has developed an entire mathematics education reform programme around it, which he calls 'computer-based maths'.[20] Wolfram argues that we live in a 'computational knowledge economy', where it's not what you know that counts so much as 'what you can compute from knowledge'.[21]

In Wolfram's curriculum, students learn mathematics through a four-step model: 1) define the question, 2) turn it into a form that can be computed, 3) compute answers and 4) interpret the results. It's a cyclical process, with one set of results sparking further questions. And because the third step of computation is automatic and instantaneous, we can experiment with multiple variants of the same question without delay. The division of cognitive labour is clear: humans ask the questions, translate them into computable form and make sense of the results, but the computing step itself is delegated to computers. Once humans have set their mind to a problem, so long as they can frame it in computable terms, a computer may well aid us en route to a solution.

There's much to welcome about Wolfram's approach: it places mathematical procedures in their proper context and tasks humans with evaluating at what point they are appropriate for a given real-world setting. It does not negate the need for students to have some familiarity and fluency with these procedures, but it also does not obsess over them like most standard curriculums do. It does, however, seem exclusively utilitarian: Wolfram adopts the pragmatist perspective that mathematics derives its usefulness solely from direct real-world applications. Yet when we direct computers to places quite removed from the physical world, they offer another benefit: satiating our deepest curiosities.

Among the most curious of all mathematical objects is the number π. As far back as at least 2000 BCE, it has been known that π is a constant – that is, the ratio of the circumference to the diameter of a circle is always the same, regardless of its size. Both your shirt button and the Earth's equator will return the exact same ratio (I am allowing the indulgence of assuming they are both perfect circles).

Approximating π has thus been a labour of love for several major civilisations. The Rhind papyrus contains a procedure that estimates π as 256/81 – around 3.16, within 1 per cent of its actual value. The Greek mathematician Archimedes made a quantum leap of progress by harnessing an iterative method that involves approximating circles with polygons that have a large number of sides. The Chinese had captured π to seven decimal places by the fifth century. And the great Indian mathematician Srinivasa Ramanujan set the pace in the early twentieth century with outrageously fanciful representations of π in terms of infinite sums.

Modern computational methods have added new thrills to the chase. In 1949, ENIAC, an early electronic computer, calculated π to over 2,000 places, almost doubling the record. On 14 March 2019,* Google announced that one of its employees, Emma Haruka Iwao, had calculated π to a staggering 31.4 trillion digits, demolishing the previous record of 22 trillion. It would take around 332,064 years to recite all 31.4 trillion of those digits. Google was the perfect hunting ground: Iawo was able to leverage its cloud-computing technologies, using 170 terabytes of data (the equivalent of around 340,000 song files)

---

*Incidentally, that date is no accident: 14 March, or 3.14, in the US month/ day format, marks the annual celebration of π Day in homage of everyone's favourite constant.

spread across 25 virtual machines. The computation lasted 121 days. The record was eclipsed in January 2020 (50 trillion digits), then again in August 2021 (62.8 trillion), with more surely to follow.

You might well ask how far we need to go. For practical purposes, not far at all: it usually suffices to approximate π to two decimal places, 3.14. Archimedes's methods took him to three decimal places, whereas Isaac Newton stopped at sixteen. NASA's Jet Propulsion Laboratory uses only fifteen digits of π when performing calculations for interplanetary navigation. For the most ardent engineer unwilling to compromise an iota of accuracy, approximating π to thirty-nine decimal places will capture the size of the Milky Way to within the length of a proton.

It is evident that efforts to capture ever more decimal places of π are driven by intrinsic motives rather than some practical end. One of the most alluring characteristics of π is that it is an irrational number, which means that its decimal expansion will never run out or repeat. As the numerical representative of the circle, π is an embodiment of all that is infinite and elusive. It can never fully be captured, which only makes chasing its tail that much more attractive. In Iawo's own words: 'There is no end with pi, I would love to try with more digits.'[22] This is Everest without its summit: an infinite ascent towards previously unchartered frontiers. And for those adamant on retaining a practical purpose, approximating π has been used as a debugging tool (if two programs produce two different approximations of π then at least one must contain an error somewhere),[23] while the Swiss team behind the 2021 record foresees applications in 'RNA analysis, simulations of fluid dynamics and textual analysis'.[24]

As machines become smarter, they can even give rise to new

types of questions. The Ramanujan Machine, for example, is a machine learning program that uses known formulae to calculate digits of π (and other mathematical constants) and then uses the first few thousand digits to predict entirely new formulae.[25] Discovering new formulae is a step beyond simply crunching through existing ones. Some formulae will be true, others false – the program is, in a sense, throwing up automated conjectures (of the kind for which Ramanujan was renowned) that humans can then get to work examining.

The pattern-matching prowess of computers is also being brought to bear in fields like topology, which often deals with complicated shapes (so complicated that we struggle to visualise them). Approaches that have proven effective in image recognition have been applied to get good guesses on what some shapes look like. For example, when an artificial neural network was given a list of mathematical knots (which are the same as everyday knots except that there are no loose ends), it confidently, and correctly, predicted that each one did not arise as a slice of a higher-dimensional object (this has a very technical definition which I won't give here). The only exception was a particular knot called the Conway knot, where the network returned a probability of ½.[26] This was uncanny because mathematicians had struggled for decades to prove that the Conway knot, like the others, was a 'non-slice'. This conjecture was later proven by graduate student Lisa Piccirillo using novel techniques, but the uncertainty of the neural network seemed to match the difficulty that human mathematicians had faced for so long in probing this particular shape. This example may signal that AI can alert us to objects worthy of our study and, far from unseating us from the helm of problem solving, direct new lines of inquiry as we take a closer look at the things machines struggle to make out.

Questions beyond computation

It is tempting to think that, with the advent of AI, computers are approaching a state of omniscience. The notion that a computer might be able to answer any question we throw at it lies at the heart of David Hilbert's *Entscheidungsproblem* ('decision problem'), which he posed in 1928. Hilbert asked whether there is a single algorithm that could determine, from a given set of axioms, whether or not any statement is provable. Gödel's incompleteness theorems would demonstrate that, within any system containing the rules of arithmetic, there would be statements that could be neither proven nor disproven, eviscerating Hilbert's earlier hope of a complete and consistent mathematical system. The *Entscheidungsproblem* took aim at all provable statements. If Hilbert's algorithm existed, one would simply feed a given statement into it; if the statement was provable, the algorithm would say so and spell out the proof. If it wasn't provable, the algorithm would determine as much. The implications for human mathematicians would be both profound and dire. The mathematician G. H. Hardy defiantly speculated: 'There is of course no [algorithm], and this is very fortunate, since if there were we should have a mechanical set of rules for all problems, and our activities as mathematicians would come to an end.'[27]

Hardy would not have to wait long for vindication, and it came courtesy of the computing pioneer Alan Turing. It was in exploring the *Entscheidungsproblem* that Turing formalised notions of computation. That's right: the foundations of modern computing were motivated by abstract mathematical questions. Once Turing had meticulously defined constructs such as *algorithms*, *programs* and *machines*, he delivered the fatal knockout to Hilbert's hope of solving all of mathematics.

If you have ever found yourself staring at a frozen computer screen, watching the infamous hourglass (or rainbow wheel if you are of the Apple persuasion) and wondering whether you should give it more time or reboot the whole system, then you might also have wondered whether a program exists that can tell you one way or another. Using similar ideas to Gödel's incompleteness theorems, Turing showed in 1936 that no single program can determine whether any other program runs forever or halts. Turing supposed the program does exist, and then wrote another program based on it – specifically, the words in the following box:

> If the program inside this box finishes, then run forever.
> If the program inside the box doesn't finish, then stop.

There are two possibilities: either the program inside the box finishes, or not. If you follow the words in the box, you should agree that if it does finish then it doesn't, and if it doesn't finish then it does. Either way we reach a contradiction (this has the flavour of the barbershop paradox which inspired Gödel's true but unprovable statement which, in turn, Turing adapted to the case of computer programs). Because we arrived at a contradiction, our initial assumption – that there is a program able to decide whether any other program runs forever or halts – must be false. As far as the termination of that hourglass is concerned, you will just have to live in a state of *unknowing*.

This is known as the 'halting problem' and it has direct consequences for mathematics. It is possible to frame the halting problem as mathematical statements about whole numbers. And since Turing showed that no single algorithm can tell you whether every other program halts, it must follow that no single algorithm can tell you whether any given mathematical statement is true. There are algorithms that can solve individual maths problems, and even entire classes of maths problems, but no single algorithm can resolve the whole of mathematics. The *Entscheidungsproblem* is undecidable, after all, and mathematicians will have to exercise ingenuity to keep coming up with new ways of solving problems since no single method will work for all of them.[28] This is the first blow to the ideal of computational omniscience.

The next blow is more consequential, and it relates to the physical limitations of computers. There is only so much matter in the universe (10^{54} kilograms' worth), which means that there is only so much computational power to go around. Even among those problems that can be solved by machines, the solution is impossible to come by because it demands too much processing for a computer to handle.

As a reluctant traveller, I dread work trips that involve visits to multiple far-flung locations. I am always seeking to minimise the total amount of travel in order to reduce the physical burden on me, as well as the travel costs incurred by my employers. Suppose I have to visit five cities dotted around England, starting and ending in the same city. What I am after is the shortest route. Assuming the distance between each pair of cities is known, it is not too difficult to conjure up an algorithm that checks all possible routes and computes the distance for each one. There are five choices for the first city, then four for the next, then three, two and one. In total, that makes $5 \times 4 \times 3 \times 2 \times 1 = 120$ possible routes. I can now check each one, rank them in terms of distance and voila, my decision is made.

Now imagine I am stateside and need to embark on a megatour around all fifty states. This time, there are $50 \times 49 \times 48 \times \ldots \times 1$ (written in shorthand as 50!) possible routes. Suffice it to say this is no lean number: it is roughly equivalent to the number 3 followed by 64 zeros. If you had the most super of supercomputers at hand, computing each route in the time it takes light to cross the width of an atom, you would still have to wait for roughly the age of the universe multiplied thousands of trillions of trillions times over for the program to return an answer. The method of listing all possible routes will not take us very far practically, because for large numbers of locations, computers are unable to keep up.

This is the *Travelling Salesman Problem* and it is part of a class of problems that are known for rapidly growing in computational toughness as the numbers involved get larger. What makes this problem interesting is that it is relatively quick to verify whether a given route falls within a specified budget. There is a difference, however, between verifying a solution and finding one. Finding your misplaced house keys is notoriously difficult (something we can all attest to, no doubt). On the other hand, *verifying* that a key I hand to you is the correct one is much simpler because you can just stick it in the lock and check for a fit.

So we can place problems into two categories. The first contains all problems that can be *solved* in a reasonable amount of time – this is the class P (which stands for polynomial time and means that the running time for the solution algorithms is proportional to some power of the size of the inputs). The second contains all problems whose solutions are easy to *verify* – the Travelling Salesman Problem and your missing house key belong to this class. It is called NP (which stands for non-deterministic polynomial time, for reasons that are slightly more technical).

The killer question at the heart of it all is whether the classes P and NP are actually one and the same.[29] It's easy to see that every problem in P is in NP – if you can solve a problem reasonably quickly, then you can certainly verify a candidate solution reasonably quickly (if nothing else, you can just solve the problem and check it matches your candidate solution). The more interesting question is whether every NP problem is in P – in other words, if we can easily verify a solution to a problem, does that mean the problem is easy to solve? If so, it would mean there is a quick way to solve the Travelling Salesman Problem; a quick way to find your missing house key every time.

This is usually termed the *P versus NP* problem, and figuring it out either way will make you a millionaire because it is of such difficulty and importance that the Clay Mathematics Institute listed it as one of its seven millennium problems, each of which comes with a $1 million reward.* Kurt Gödel – he of the incompleteness theorems – played a major hand in the *P versus NP* affair. It was his 1956 letter to the legendary polymath John von Neumann that sparked research into the topic. Gödel's incompleteness theorem had already shown that mathematics cannot solve all problems: some statements are unprovable. If one could now show that there are problems in NP but not in P, it would mean that even among the problems we can solve, some of those solutions cannot be found *quickly* enough to be of practical use.

The answer to whether P and NP are identical (P = NP) could be a world-changer, depending on which way it swings. If these two classes are equal, then a whole raft of problems that we presume to be complex will suddenly submit to algorithms that can be practically applied. That would be mixed news for the world. On the positive side, it could accelerate

---

*At the time of writing, only one of the others, the Poincaré conjecture, has been solved (in 2003), although Grigoriy Perelman, the Russian mathematician behind the proof, famously declined the prize.

cancer treatments and kidney exchanges, revamp forensics and bring untold logistical savings to companies, to name just a few applications. At the same time, it would spell disaster for cybersecurity. The privacy of your online banking data relies on the difficulty of breaking really large numbers down into their prime factors – a problem known to be in the NP class (if I give you two prime numbers, you can quickly multiply them to see if the result equals a given target). A world in which P = NP would suddenly render your most secure data vulnerable to hacks. Mathematicians themselves might not be spared from the fallout, a point not lost on Gödel when he declared that if P = NP, then 'the mental work of a mathematician concerning Yes-or-No questions could be completely replaced by a machine'.

The majority view is that P does not equal NP after all, which would signal business as usual. The advent of quantum computing, which allows for exponentially more calculations to be executed, might shift the landscape by creating new categories of complexity,[30] though the overall picture is likely to remain the same: we will need more creative and holistic approaches to solving those NP problems because brute force won't get us there in good time.

Mathematics thrives on open-ended problems, and the existence of problems that are easy to verify but hard to solve reshapes these clean-cut problems into messier ones. Entire fields of mathematics and computer science are dedicated to finding *efficient* and *approximate* answers to problems whose solutions escape our precise calculating tools. Even while exact, deterministic solutions remain elusive, approximations often suffice – after all, does your company really need the absolute, cheapest travel route, or just one that is reasonably cheap?

The *P versus NP* problem sets up a cat-and-mouse race

between our ability as humans to dream up questions, and the power of computers to answer them. If it does transpire that some NP problems fall outside the category of P, it will speak volumes about how far our minds can wander – far enough, evidently, to transcend the computational might of machines.

Let's go back to Sudoku. The standard format of nine characters is easy prey for most computers. It is an NP problem because one can easily verify candidate solutions – just check the rows, columns and boxes. It is also in the P class because there are few enough permutations to enable even brute force techniques to get the job done in reasonable time. Things get more interesting when we expand Sudoku to, say, a 25 × 25 grid. The rules are the same as before, except this time every row, column and box needs to accommodate each of the numbers 1–25. We are still in the class NP, but this time the computer will hum away seemingly indefinitely as it works through an impenetrable number of possibilities. This version of *mega-Sudoku* is possibly, and quite probably, outside class P. The standard version of Sudoku is still good conditioning for our minds, but the leap towards larger versions of the same game that defy brute-force processing is the real triumph of human thinking, because it signifies that the reach of our deep-rooted curiosities can outstretch the prowess of machines.

Another twist in the mega-Sudoku tale is that it belongs to a more specific class of problems within NP that are said to be NP-complete (the Travelling Salesman Problem is also NP-complete). This class is so named because all NP problems can be reduced to these problems in reasonable time. If we solved a single one of these problems in a reasonable time, then we could solve every NP problem in a reasonable time, thereby showing that P = NP after all. By mastering those large Sudoku

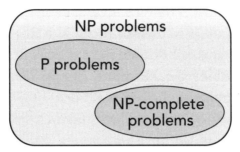

A graphical representation of the P, NP and NP-complete classes. If a single NP-complete problem is shown to be in the class P, then all three classes are actually the same.

grids you will, as a by-product, have figured out how to solve all those other NP problems quickly.

Giant Sudoku isn't likely to yield to computers any time soon, and nor is mathematics. And even if they do, humans can breathe easy. As mathematician Jordan Ellenberg puts it: 'We are very good at figuring out things that computers can't do. If we were to imagine a future in which all the theorems we currently know about could be proven on a computer, we would just figure out other things that a computer can't solve, and that would become "mathematics".'[31]

As computers march towards the frontiers of new knowledge, the challenge for humans will be to manage all that complexity. If computers only ever amount to solution hunting, then there is still a role for us humans to decide which questions are the most interesting, which should be the preserve of humans, and which need expanding. Computers can be a wonderful aid alongside our explorations, but it is we humans who chart the journey.

Reviving our childhood habits

Neoteny is the wonderful term that describes how adults retain some of their juvenile features. Somewhere along the path to adulthood, we seem to lose our tenacity for questioning. Formal schooling, where the focus is squarely on getting the right answer to prescribed questions, has much to answer for. The social critic Neil Postman recognised this decades ago when he posed his own (albeit rhetorical) question: 'Is it not curious, then, that the most significant intellectual skill available to human beings is not taught in schools?'[32] The twentieth-century Brazilian educationalist and activist Paolo Freire was more damning still, employing a banking metaphor to describe how teachers are too often expected to 'deposit' knowledge into the minds of students who, for their part, are expected to passively receive, store and file these deposits.[33] For the activist Freire, this amounts to an 'ideology of oppression'. Preventing students from asking questions and stifling their lines of enquiry is, according to Freire, an act of 'violence' that 'alienates human beings from their own decision-making'. It strips them of their agency to think for themselves, to generate their own questions, and to develop a worldview of their own choosing.

The good news for adults is that we can reorient ourselves to our curious, child-like selves, although sometimes it takes some nudging. Research has shown that people can learn to 'think young'. In one study, a creativity test was issued to two groups of adults. The adults in the first group were encouraged to think of themselves as 'seven-year-olds, enjoying a day off from school', while the other group defaulted to thinking of themselves as adults.[34] With that encouragement, the first group came up with more original ideas and displayed 'more flexible, fluid thinking'. In the business world, the analogue of a *curiosity quotient* (CQ) is challenging more traditional measures of intelligence such as IQ, with employers seeking out people who 'find novelty exciting'[35] – how they would have loved to have Martin Gardner on their books.

Progressive educators like to emphasise the distinction between *consuming* knowledge and *creating* it. This effort can begin at school by placing questioning at the heart of the learning experience. Freire called for 'problem-posing education' in which students and teachers usher one another, through mutual probing and reflection, towards 'a constant unveiling of reality'. The concept is hardly new; philosophers as far back as Socrates advocated teaching based on dialogue rather than lecturing, where the express aim is to equip learners with the power to think critically, which means (for one thing) thinking for themselves. Socratic dialogue is premised on tenacious questioning that forces students to reflect deeply about the knowledge they possess.

As AI strengthens its foothold on our everyday lives, the distinction between consumption and creation matters to us all. Freire's banking model fits squarely with the computational paradigm of school maths, and the alternative he sought is a lodestar for the type of mathematics portrayed in this book. We can choose to be subjects of AI, placing ourselves at the mercy of algorithmic, automated decision-making and consuming these technologies without scrutiny, or we can demand to know how these technologies work, what risks they carry, what threats to equality and justice they pose. We can choose, in other words, to be co-creators of the so-called 'machine age' by fiercely holding on to our curious instincts and never surrendering our ability to ask questions of how AI can be designed to serve our human goals.

This is why the recreational flavour of mathematics matters

so much. Puzzles and problems are not only gratifying to solve; they also sharpen our ability to *ask* interesting questions in the first place. Even the most frivolous questions – like whether it's possible to trace a particular path around a city or how to divide the spoils of an unfinished game – can spawn entire fields as we thread together their common elements. The field of AI (including its many subfields) is itself still nascent, and its development very much depends on the questions we choose to pose. It is vitally important that we reject any deterministic notions of what the field *must* become, recognising that humans have agency and therefore the power to shape the ways in which these technologies are designed and implemented. When we evaluate the effectiveness of a prediction model, for instance, we may ask questions about its overall accuracy. Or we may instead focus on where our model errs, probe the consequences of its mistakes, and ask who is affected the most, and what the trade-offs are between automation and fairness. The models we end up with, the conclusions we draw and the impact they have on our society are informed very much by our questions.[36]

Computers can be allies to curious-minded humans by extending and proliferating the questions we pose. We've already seen how *Nikoli* relies on its community to devise new and interesting types of grid-like puzzles, and then outsources the task of producing several instances of each puzzle to a computer. Computers can even inform our judgements of what puzzles and games count as 'interesting'. Take the rules of chess, which have evolved several times throughout its 1,500-year history. For instance, it was only in the 1400s that the queen was granted the ability to travel multiple squares in any direction. The same DeepMind technologies that have mastered games like chess and Go are now being called on to explore alternative rulesets. It is easy enough to task these programs with playing out millions of instances of chess under different rules. We can then evaluate game patterns to see which combination of rules leads to the most exciting gameplay dynamics.[37]

But there are some questions that computers cannot answer. This chapter has, through the lens of mathematical problems, examined the fundamental limits of computers, as well as some practical ones. The messy realities of the physical world may be another inhibiting factor in the quest for machine omniscience. For self-driving cars to go mainstream, for instance, we need to extract insights from philosophy and ethics to understand what choices an algorithm ought to make when confronted with fatal scenarios. Philosophers have mulled over thought experiments like the 'trolley problem', which asks when it is justified to sacrifice one group of people over another (typically larger) group. But there are few definitive answers to be sought here: the toughest questions deal in shades of grey. Puzzles involving life and death, ethics, morality, religion and the law have confounded humans for thousands of years; they're unlikely to succumb to machines any time soon. Some questions evade the precise, binary language of computers, and others do not lend themselves to answers a computer can make meaningful. The danger is that we alter – maybe even dilute – our questions to a form that computers can process, and blindly accept the answers.

The muddiest questions are so often the ones that matter the most. They invite us to reflect on our worldviews, to examine our core beliefs and values, to entertain ambiguity and uncertainty. Sometimes an answer is the worst thing one can demand of a computer.

PART II

WAYS OF WORKING

6

TEMPERAMENT

Why speed is overrated, getting into flow,
and the wisdom of 'sleeping on it'

In a timed battle of calculational wits between humans and computers, few among us would dare to step up. Art Benjamin is an exception. The self-styled 'mathemagician' performs stupendous feats with numbers, rarely leaving his audience disappointed. In one TEDx talk, you can see Benjamin pitting himself against a calculator by multiplying numbers in his head at rapid speed.[1] Sure enough, he can multiply a pair of two-digit numbers before his volunteers can type them into their pocket calculators. But why stop there? Benjamin takes to squaring numbers of increasing size, culminating in his show-piece: the square of a five-digit number. To add to the intrigue, Benjamin relays his thinking out loud, murmuring mysterious phrases such as 'light tomorrow', that presumably anchor each step of his calculation. True to form, Benjamin rattles off the digits in succession, all correct, takes his bow and departs triumphantly. The attentive viewer will notice, however, that Benjamin has stopped racing by the end – he knows that even a modest calculator will win out against his more complex end-of-show stunts. And if Benjamin was up against IBM's Summit supercomputer,[2] which performs 200 quadrillion calculations per second, his act would not even achieve lift-off.

The digital age is steeped in a narrative of rapid change.

Humans are encouraged to keep up by speeding up. Yet when it comes to conscious processing tasks like calculation, we have no hope of keeping pace with computers. On the trade floor, for instance, computers execute trades within microseconds, a pace no human – not even Art Benjamin – could hope to match. The human brain needs a quarter of a second simply to respond to a stimulus, let alone to process on-screen information and click the relevant buttons to buy or sell assets. The frantic pace of automated calculation gives us no time to analyse its potential side effects. In the 2010 Flash Crash, investors saw nearly $1 trillion of value wiped off from US stocks within 36 minutes.[3] Speculation and analysis followed as to how such a volatile dip could come about. It took several months for US regulators to point the finger at a sell order instigated by the mutual fund Waddell & Reed. At 2.32 p.m. on the day of the crash, the fund triggered an automated algorithm-trading strategy to sell contracts known as e-minis. It marked the biggest shift in any investor's daily position that year, and set in motion a selling frenzy among traders. The market rebounded and closed within 3 per cent of where it started the day, but the episode is a warning shot to those who place faith in algorithms on account of speed alone.

The brain-as-computer metaphor seems to flatter us humans unduly – we can get as far as processing one short sentence at a time (around 40–60 bits a second), but that's it. *If only* we could calculate at the speed of computers. In fact, depending on the type of task, humans more than hold their own. We are phenomenally adept at processing the world around us, a task that comes so naturally to us that it largely escapes our conscious attention. When we take in a scene, we may be aware of certain details like the relative luminance of objects, but others – like the precise amount of overall light level – are dealt with unconsciously.

The brain relies on thousands of processors working in parallel, each propagating data across millions of nerve fibres within the brain. The retina alone (which, pushing the metaphor further, can be thought as the brain's 'webcam') packs in 100 million neurons into a 0.5mm-thick square centimetre, allowing it to process ten one-million-point images per second.[4] Estimates of the brain's overall processing power sit at around a petaflop, or a thousand trillion operations per second.[5] By that measure, brain-as-supercomputer is a more appropriate metaphor.[6]

The brain manages all this with an average energy consumption rate of merely 12 watts; by comparison, your laptop consumes around 100 watts. The brain is built for such extraordinary versatility and efficiency that it defies comparison with brute-force processing devices.

Computers are relatively simple beings. We can scarcely tolerate slow ones: faster means better, no exceptions.[7] In addition, there is one metric that software developers value above all when evaluating their systems: *uptime*. This relates to the percentage of time a system is live (or 'awake'), and developers will expend immense effort to inch this metric towards the ideal of 100 per cent. Uptime reflects the implicit recognition that, as yet, there is no unconscious layer to computers, no movement between different stages of alertness. For humans, the picture is rather more complicated. A recurring theme of the previous chapters is that humans possess multiple modes of thinking. We have an approximate sense of number as well as a facility for precise calculation. We can reason through problems slowly and methodically, but we also possess rapid-fire intuitions and impulses. Humans think fast and slow, consciously and unconsciously, and many mysterious shades in between.

If your brain was a company, then its headquarters would

be situated at the frontmost lobe of each hemisphere. Known as the *prefrontal cortex*, this part of the brain is found only in mammals and, for humans, it is the command-and-control centre for our behaviour. Without the prefrontal cortex, we would be left at the mercy of our automatic responses to environmental cues. It is one of the main brain regions responsible for supervising our thoughts, planning our actions, making decisions and detecting errors when we deviate from our goals – all of which is collectively referred to as our 'executive function'. It all seems very orderly, ideas managed from the top down. But before your executive function can, well, function, it needs some way of grabbing hold of ideas in the first place. A mesh of thoughts revolves deep within our subconscious. With each firing of a neuron, ideas form and compete with one another for our conscious attention. We're only just starting to grasp how this competition plays out, and how the most novel ideas rise up through our cognitive filters to the forefront of our minds.

Humans, unlike computers, have metacognitive awareness – we can think about how we think and regulate our own mental behaviours to extract maximum output from those 12 watts. To produce our most creative work, to solve our most stubborn problems, and to keep high-pace computers in check, we must craft an alternative narrative for the digital age, one that embraces multiple modes of thinking. We must learn when to privilege patience and restraint over speed and acknowledge 'downtime' as an essential feature of our brains. This ability to self-reflect and to fine-tune our ways of thinking is what I refer to as *temperament*. And it carries special significance for mathematics, a discipline where competence is too often conflated with speed.

Breaking the cult of quickness

We are enamoured of speed, not least when it comes to feats of calculation. Art Benjamin is revered because he is fast with numbers, and in some quarters, mental maths attracts a cult-like following. It is the brand of mathematics that promises its loyal subjects the honourable label of mathematical *genius*.[8]

It also makes for good television. The UK's search for its ultimate *Child Genius*[9] has taken on the form of a televised contest, complete with bloated pep talks from expectant parents and high-octane drama as kids take to the stage. The mathematics round has a predictably blunt format: timed arithmetic problems that often reduce contestants to tears. In 2008 I had my own brush with timed challenges as a series winner of long-running UK gameshow *Countdown*, where we were given thirty seconds per round to solve anagram and arithmetic problems with the famous clock (not to mention its annoyingly repetitive tune, *ba-da ba-da ba-da-da-dum boom!*) bearing down on us as the seconds ticked by.[10] Riveting as the experience was, it was the worst possible advert for my mathematical skills, reinforcing a perception among my friends and family that I spend my working days immersed in complicated sums.

Our fixation with high-speed sums can be seen in the prevalence of rapid-fire mental maths systems across the world. In India, there has been an explosion of interest in the number play of Vedic Mathematics, following the publication of Sri Bharati Krsna Tirthaji's 1965 book of the same name. A cottage industry of course-offerings promises to teach students 'perhaps the most refined and efficient mathematical system possible'.[11] It's a bold claim, yet the content appears to be no more than variations on a common theme – contrived mental gymnastics applied to select calculations, with emphasis on speed. Among other arbitrary titbits, you'll find procedures for computing square roots to nineteen decimal places. Tirthaji claimed that his methods were derived from sixteen word-formulae, or *sutras*, that have their origins in ancient Hindu scripture (claims that have been roundly debunked).[12] Such methods do a great disservice to the rich and multifaceted mathematics found in the ancient Vedic tradition. Among a multitude of mathematical explorations, Vedic texts include early gestures towards right-angled triangles (what would later be known as Pythagoras's theorem) and geometric approximations to squaring the circle (also credited to the Greeks)[13] – mathematics far more profound than Tirthaji's arithmetical tricks.

The Trachtenberg system is cut from similar cloth.[14] It is named after Jakow Trachtenberg, a Russian Jewish engineer who developed the methods while being held in a Nazi concentration camp, in an attempt to keep his mind occupied. Sadly, the remarkable origins of the system cannot rescue its methods from the trappings of convoluted abstraction: it may well take a professional engineer to grasp their significance. Or it may take a fictitious seven-year-old like Mary Adler, the *Genius* portrayed by Mckenna Grace in the film of the same name, in which the child prodigy draws on the Trachtenberg system to demonstrate her flair for numbers.

These systems may have held currency in the era of human computers when it literally paid to do your sums quickly. Now that the arc of technology has bent towards automation, and human computers have been banished from the labour force, systems like Vedic and Trachtenberg, mathemagicians like Benjamin, even *Countdown* winners, should be celebrated for their quirks but nothing more. As a conceptual system, mental arithmetic does relatively little to elucidate the principles of mathematical intelligence. At worst it perpetuates the falsehood that mathematical intelligence is a function of speed; a view unfit to be applied to humans.

This is not to say that speed should be shunned altogether. To master any craft, we need to acquire fluency of its most basic elements. My most painful learning experiences were behind the driver's wheel. It seemed to take forever to me to get to grips (literally and figuratively) with the gear lever. With every gear change, there was so much to keep track of: my current speed, the patch of road ahead, the pressure of my right foot on the accelerator, the placement of each gear number marked on the lever handle. Only after several hours of practice (more than I care to admit) did I acquire the familiarity to action the gear changes with minimum conscious effort. With the automaticity of these basic skills shored up, I could direct my attention to all the other subtleties of driving.

This need to 'free up' our attention applies to everyone. The field of cognitive psychology has revealed an important feature of the human brain. Broadly speaking, our brains manage ideas in two forms: long-term memory and working memory.* Long-term memory relates to ideas that are embedded in our subconscious and can be recalled at will – it's how you read these sentences so effortlessly, devoting almost no conscious attention to the individual letters that make up each word. Working memory is quite the opposite; it is very closely related to executive function and speaks to the conscious aspect of our thinking that allows us to manipulate information. Working memory is the mental sticky note for short-term problem solving, and what makes it so significant is its modest

---

*An important caveat is not to take this description literally. Memories are not physically stored in discrete repositories of our brain as they are in computers: they exist as distributed representations.

size: humans can only host, at most, four to seven conscious thoughts at a time (no wonder I felt overwhelmed at the wheel in those early stages). The amount of *cognitive load* our brains can handle at any time is limited, which is why we marvel at those who can manage multistep calculations in their heads so effortlessly. Experts in all domains have, through huge amounts of practice, reconfigured their neural connections so that the process behind their skill feels familiar and mundane – unworthy of conscious attention and therefore in no need of the efforts of working memory. This is the 'muscle memory' you so often hear about with regard to motor skills.

The cognitive load perspective is used by many educators to justify cramming all those procedures into students' minds. If the goal is to free our minds to focus on the complex aspects of problems, then sinking facts and methods deep into the trenches of our long-term memory so that they can be summoned at will may have renewed purpose. The reason we were asked to memorise all those pesky multiplication facts at school is that if we expend any effort calculating them on the fly, our working memory will quickly become fully occupied and unable to accommodate deeper ideas.

It is tempting to think that with digital calculators at hand, we can consider abdicating altogether. What better way to free our minds, the reasoning goes, than to offload the burden of calculation to computers? But that will not do much to relieve our cognitive load because entering numbers into a calculator still requires some conscious effort. This is another reminder of the irony of automation that recurs throughout this book: to hold machines to account, we need to engage with their core competencies.[15] Some level of calculational proficiency is prudent.

We need to be wary, however, of extreme deference to

speed: when all learning is reduced to the consumption and rapid recall of individual facts, other aspects of intelligence – such as the five principles covered in previous chapters – are often left on the margins. When we are expected to serve up answers automatically, we often grab the first one that comes to mind, without even a moment of reflection. We've seen how thoughtless calculation is a recipe for nonsensical answers. We must therefore retain some conscious awareness of our mental calculations. Speed should never come at the expense of having a good sense of number, being able to create diverse representations of concepts and reasoning through arguments. In fact, speed can emerge as a by-product of a more flexible understanding of how certain facts and procedures relate to one another.

The side effects of high-speed mathematics

As quickly as you can, answer the following three questions:

A bat and ball cost £1.10 in total. The bat costs £1.00 more than the ball. How much does the ball cost?

In a lake, there is a patch of lily pads. Every day, the patch doubles in size. If it takes 48 days for the patch to cover the entire lake, how long would it take for the patch to cover half of the lake?

Suppose your doctor offers you a new test for a rare disease. The disease afflicts around 2.5 per cent of the population and the test is 80 per cent accurate. Being the cautious person that you are, you take the test. Bad news: you have tested positive for the disease. Based on this information, how likely are you to have the disease?

High-speed mathematics, the kind you've just engaged in, should come with a hazard warning for all its unintended consequences. Chapter 3 covered the first of these: *cognitive bias*. As we saw there, thinking fast (or 'System 1' thinking, as opposed to the slow thinking of 'System 2') is a fertiliser for logical discrepancies, especially when dealing with subtle truths. Mathematics is replete with subtlety, which means it's often a bad idea to race against the clock for problems that require even a modicum of reasoning.

The first two problems appear in the Cognitive Reflection Test, which was the basis of a 2005 psychological study showing that people tend to solve problems without reflecting very much on the details.[16] When pressed for an answer, many people (perhaps even you?) opt for 10 pence and 24 days, respectively. The correct answers, in fact, are 5 pence and 47 days, both of which are easily deducible with some considered reasoning or a momentary reflection of one's first response.

The third problem is among the many surprising truths involving probabilities. The answer is just over 9 per cent, far lower than most of us would guess (we tend to ignore the fact that the disease is so rare to begin with that even a positive test is not cause for alarm – a cognitive bias known as 'base rate neglect'). Probability is a topic that perpetually leads humans towards System 1 biases. Its truths often run contrary to our immediate perceptions – educator and writer Sunil Singh terms it the 'devil's mathematics' for this very reason.[17]

As with other cognitive biases, knowledge and intelligence only go so far in rescuing us from these thinking errors; more than half of Harvard and MIT graduates slip up on the bat/ball problem and, rather worryingly, over 85 per cent of healthcare professionals are unable to solve the diagnosis problem.[18]

Yet the bat/ball problem is, once properly considered, a relatively straightforward arithmetic problem that requires some modelling of subtraction. The lake problem is a basic application of exponential growth. And the probability riddle relies on Bayes' Theorem, a formula for calculating the likelihood of one event given information about another. Mathematics equips us with the tools to solve a wide cadre of problems, but its potential goes to waste when we lunge for immediate answers. The human brain does not readily intuit concepts like exact calculation, exponential growth or Bayes' Theorem.* The way to exploit these acquired models of the world, and to silence our faultiest intuitions, is to slow down our thinking processes. The mathematician Ian Stewart advises: 'The most important thing about probability is not to intuit it.'[19] In other words, allow careful, deliberate reasoning to steer your thoughts. Stewart's advice extends to most corners of mathematics: the most effective shield we have against System 1 errors is our ability to slow down.

This should come as welcome guidance to anyone who has suffered from *maths anxiety*, the technical diagnosis given to people who approach the subject with dread. The ailment is more pronounced in mathematics than in any other subject, and it is not even limited to low-attaining students – one study suggests that over three quarters of maths-anxious students

---

*There's an important distinction here between *Bayesian reasoning* – updating beliefs based on new information – and the precise formulation of Bayes' Theorem, which results in probability estimates. We use Bayesian reasoning all the time in a loose sense, but struggle to intuit the actual likelihood of certain events occurring.

are 'normal to high achievers' in school.[20] Our obsession with speed is a major contributor to maths anxiety. When mathematics is reduced to a timed performance act, it becomes a high-stakes competitive affair where speed is synonymous with rank and status. In school, maths drills that are premised on speed and accuracy are often the root of people's detachment from the subject. There is a cruel irony to drills. They are intended as a means to embed facts in our long-term memory, which, as we have seen, frees up our limited working memory to think through and solve problems. Yet the stress induced by drills clogs up that same working memory. The amygdala, an almond-shaped cluster of nuclei found deep within the brain's temporal lobe, acts as an emotional filter, directing sensory inputs to different parts of the brains for processing. When thoughts of impending failure rush in, the amygdala instead directs these inputs to the reactive 'fight, flight or freeze' regions of the brain. Moreover, when the brain is stressed it produces cortisol, which invades our hippocampus, the gateway through which information must pass to be memorised.

The net effect of the brain's stress-response mechanisms is that those limited slots of conscious thinking quickly fill up, leaving our minds with little room to process the problem at hand. In the most severe cases, we may be often overtaken by a *choking* sensation, a term used in sports and formalised by psychologist Sian Beilock.[21] Choking paralyses our ability for productive thought by steeping learning in the fear of failure.

As well as needlessly inducing fear, high-speed solution grabbing distorts mathematics into a form unrecognisable to professional mathematicians. Just as fast-paced forms of chess diminish the importance of more reflective gameplay, the essence of mathematical intelligence is lost when we reduce the subject to rapid-fire questions and answers. Problem solving takes on an entirely new flavour when it is restricted by time: it reduces mathematical intelligence to the retrieval and execution of familiar techniques, which by itself does not lead us to the mathematical vistas of previous chapters.

What the past few chapters have shown is that, for humans, knowledge is as deep as it is connected, as open-ended as it is strictly procedural. When we retrofit the human brain into blunt processing machines, we strip mathematics of its character as an exploratory, sense-making discipline.

Mathematicians have no qualms with taking their time. The late Maryam Mirzakhani was proud to admit she was a slow thinker who was attracted to deep problems that she could chew on for years. 'Months or years later, you see very different aspects' of a problem, as she said.[22] There are problems she thought about for more than a decade without finding the answer. As another mathematician (and Fields medallist), Timothy Gowers, put it: 'The most profound contributions to mathematics are often made by tortoises rather than hares.'[23] For the deepest and most rewarding problems in mathematics, slowly does it.

Switching off

Computers don't suffer from maths anxiety. Information overload barely registers as a concern, save for problems that require unfathomable amounts of computation. Slowing down is counterproductive because it will make no difference to how computers process information or solve problems. The biological quirks of the human brain that we must endeavour so hard to compensate for don't seem to apply to computers. By the same token, computers miss out on the benefits that a drastic change in thinking speed confers on us. For humans, slowing down isn't just a way of guarding against our anxieties and biases – it can also pave the way for our most creative feats.

The occasional thought hits us as a flash of insight – that moment when revelation dawns and an idea that previously seemed impenetrable suddenly falls into place. Moments of ingenuity are not simply blind luck so much as states that can be engineered.

Mathematicians have long suspected that there is a mysterious element to creative problem solving. The French mathematician Henri Poincaré described his creative thought processes in terms of choice:

> To invent, I have said, is to choose; but the word is perhaps not wholly exact. It makes one think of a purchaser before whom are displayed a large number of samples, and who examines them, one after the other, to make a choice. Here the samples would be so numerous that a whole lifetime would not suffice to examine them. This is not the actual state of things. The sterile combinations do not even present themselves to the mind of the inventor.[24]

There are infinitely many ways to glue together bits of information, some more useful and interesting than others. Creative thinking results from extracting only the most salient and sometimes the most unexpected links between what we know to reach new insights. This is not a role the conscious mind alone can fulfil – as Poincaré goes on to say: 'The role of this unconscious work in mathematical invention appears to me incontestable.'

Creative thinkers so often pay tribute to the power of unconscious thinking.[25] For the graphic designer Paula Scher, creative thought is akin to a slot machine that organises jumbled thought into a coherent sequence. For T. S. Eliot, the mind of the poet transforms fragmented thoughts into beautiful ideas. And the German polymath Gottfried Leibniz spoke of music's pleasure in terms of 'unconscious counting'.

Elaborating on Poincaré's reflections, another French mathematician, Jacques Hadamard, spoke of four stages of problem solving that flit between the conscious and unconscious.[26] First there is the conscious effort of *preparing* the mind. Next comes *incubation*, where our unconscious mechanisms get to work, seeking out new and novel connections between ideas. Most of your unconscious thoughts remain buried, but the occasional spark will percolate back towards your consciousness – what Hadamard called *illumination*. Finally, there is another conscious step in *verifying* your new insight. In simple terms: if we make the conscious effort to ask interesting questions, then we can trust our brains to undertake the unconscious effort of finding hard answers. Responding to Hadamard in a letter, one Albert Einstein pointed to the 'combinatory play' that is 'the essential feature in productive thought', concluding: 'It seems to me that what you call full consciousness is a limit case which can never be fully accomplished.'[27]

It would take a brave person to bet against Poincaré, Hademard and Einstein all at once, and their formulations are being vindicated by neuroscience. The emerging view is that when we are faced with a problem (mathematical or otherwise), our brains sort through different candidates. Somewhere outside your consciousness, the left and right hemispheres of your brain generate ideas that compete for your awareness.[28] It is thought that your left brain reaches for the most obvious associations, whereas your right brain goes on the hunt for more novel solutions. Your brain needs some judging mechanism to mediate between your two hemispheres and decide which ideas, among the obvious and less obvious, should be elevated to your consciousness. The anterior cingulate cortex, a collar-shaped region that lies beneath the cerebral cortex, is among the parts of the brain responsible for this role.

One direct way to engage our unconscious is through sleep. The Hungarian mathematician George Polya advised mathematics students to 'take counsel of your pillow' when caught in the web of a problem.[29] The psychologist Howard Gruber extended this advice by encouraging creative thinkers to make use of the three Bs of Bed, Bus and Bath.[30] Each of them relaxes the mind, enabling it to switch off from problems and allow novel connections to form deep in our subconscious layers of thought. Thomas Edison applied these ideas in the most deliberate manner. He is known to have turned power napping into a craft by seeking a sweet spot between conscious and unconscious states, from which he believed his deepest insights would emerge. His method: to hold a bunch of ball-bearings so that they would clatter onto the floor, waking him up at the opportune moment, just before he descended into full-blown sleep.

There is a strong neurological basis to the 'sleep on it' adage.[31] Even as our bodies rest during those precious hours of shut-eye, our brains remain active by attempting to replay events from the previous day and converting them to memories. We experience more than we can possibly remember, and two brain-wave functions actively sort through all the neural circuits we activated the previous day: one strengthens certain memories, the other prunes the remaining candidates. How much we remember is correlated with the amount and quality of sleep. In terms of the *types* of memories we gain, deep sleep helps us to consolidate knowledge (sometimes called 'declarative knowledge'), while the more wakeful Rapid Eye Movement

(REM) phase reinforces routines and motor skills ('procedural memory'). It is also predominantly during the REM phase of sleep that ideas float around in our brains, and the most novel connections among them are discovered.

The link between sleep and learning is so pervasive that the phrase 'sleep on it' exists in most languages. I am always bemused by accounts of high-powered productivity gurus who claim to forgo sleep and dish out a list of all the things they accomplish before some ungodly morning hour. It turns out that sometimes the most useful thing you can consciously do before 6 a.m. is nothing at all. As some philosophers have noted (contrary to biblical wisdom), 'seek *not* and thou wilt find'.

'Sleep on it' is often the only wisdom I have to offer my students, and to myself, when we are in search of inspiration. We have all experienced the 'tip of the tongue' phenomenon – the name we cannot quite recall despite intense, conscious effort, which then flashes back into our minds when we're least expecting it. Various psychological studies attest to the fact that, tempting as it is to plough through a problem or puzzle whose solution eludes us, sometimes the best course of action is to allow for a mental impasse. An idle mind will give way to insight more often than you might think.[32]

The creative mind is one that astutely navigates between conscious and unconscious states. The first allows us to focus intently on tasks, while the second gives us time to think freely, even to the point of switching off. The key is to alternate between these stages of immersion and reflection, so that ideas can first take root and then swirl around as we forge new connections.

There is an indescribable joy, or at least relief, that comes with discoveries that arise in this way. 'It is like a kind of grace,' says statistics professor Thomas Royen. 'We can work for a long time on a problem and suddenly an angel – [which] stands here poetically for the mysteries of our neurons – brings a good idea.'[33]

Embracing the struggle

No mathematical tale is complete without some recounting of struggle. Andrew Wiles knows more than most about struggle, having obsessively dedicated his career to solving a problem many believed was beyond reach. In proving Fermat's Last Theorem,* Wiles created new branches of mathematics and made connections between fields in ways never conceived. His secret:

> What you have to handle when you start doing mathematics as an older child or as an adult is accepting the state of being stuck. People don't get used to that. They find it very stressful … But being stuck isn't failure. It's part of the process … It's not that we're [mathematicians] any different from someone who struggles with maths problems in third grade … We're just prepared to handle that struggle on a much larger scale. We've built up resistance to those setbacks.[34]

Cedric Villani has written a real-time account of how mathematicians sweat and struggle before arriving at their deepest insights,[35] charting his quest to prove his own major theorem, which ultimately landed him the coveted Fields Medal in

---

*The theorem states that there are no non-zero whole numbers x, y, z satisfying the equation $x^n + y^n = z^n$ for any whole-number power n greater than 2. A proof had eluded mathematicians for over 350 years.

2014. It is a wonderful juxtaposition of complex mathematics (Villani is not afraid to share snippets of his research) and a deeply human struggle to find his elusive breakthrough. You will find Villani pacing in the dark, as well as exchanging nervous, sometimes resigned emails with his collaborator as they both contend with the possibility of failure. In a shorter reflection, the mathematician Silvia Serfaty uses the metaphor of a hike to describe mathematical research – for her, frustration is baked into mathematics, but the view from atop a solved maths problem is well worth the sweat you spill to get there.[36]

There is an entire literature centred on coping mechanisms for struggle, rooted in the psychology of how we learn. Psychologist Carol Dweck has popularised the notion of a *growth mindset*: the belief that intelligence is fluid and very much in our own control.[37] It stands in opposition to the belief that intelligence is immutable – the *fixed mindset*. Over three decades of research, Dweck has shown that a growth mindset leads to improved performance across all walks of life, from students' test scores to athletes' performance in the heat of competition. A related concept is *grit*, defined as 'the tendency to sustain interest in and effort toward very long-term goals'.[38] Grit is about persisting after setbacks, and while the research behind it is not as developed as that for mindset, there is evidence that this quality, too, is a predictor of academic and other life outcomes.

These perspectives from psychology attach neatly to our emerging understanding of how the brain works. The ability to cede conscious control of problems and place faith in our unseen thought processes is closely tied to psychological traits like mindset and grit. We're more likely to slow down and switch off from problems if we believe in our capacity to grow and find connections that have eluded us hitherto. Struggle is a fertiliser of novelty and insight because it allows room for our unconscious thought mechanisms to take hold and for our most original ideas to surface.

A growth mindset also reminds us of our *neuroplasticity*. Learning, in the end, boils down to rewiring our brain structures: creating neurons, strengthening their synaptic connections, creating new pathways and pruning unused ones. Recall the enlarged hippocampi of London taxi drivers who have devoted hours of study to memorising city routes. To have a growth mindset is to embrace the notion that we are architects of our own brains.

Computers do not have the same freedom to rewire themselves. Today's artificial neural networks are based on the idea of increasing or decreasing the strength of connections between neurons. The idea of pruning them, or growing entirely new ones, is alien. What's more, if they are guided by a fatally flawed model or procedure, then even in the best-case scenario a solution may prove elusive (they're often trapped in a so-called 'local optimum'). Switching off will make no difference because computers do not learn or develop in any way during their downtime; as soon as they are switched on they will resume towards their dead ends. The only way to rescue them from their frustrated efforts is to rethink the models they are based on – the need for human intervention is inescapable, and it may require a good night's sleep on our part to make the decisive shift.

When maths becomes addictive

The Greek scientist Archimedes is best remembered for an incident involving a scientific revelation, a bathtub and spontaneous public nudity. The *Eureka!* moment gives us a snapshot into the minutiae of Archimedes's daily existence. What's more interesting still, and perhaps more revealing of Archimedes's orientation towards mathematics, is how he died. The historian Plutarch describes the fateful scene in his account of the Roman siege of Syracuse in 212 BCE:

> He was by himself, working out some problem with the aid of a diagram, and having fixed his thoughts and his eyes as well upon the matter of his study, he was not aware of the incursion of the Romans or of the capture of the city. Suddenly a soldier came upon him and ordered him to go with him to Marcellus. This Archimedes refused to do until he had worked out his problem and established his demonstration, whereupon the soldier flew into a passion, drew his sword, and dispatched him.[39]

Archimedes is situated in a long line of humans who have found themselves caught in the grip of a problem. The consequences are not always so drastic,* but invariably the problem solver loses all awareness of their surroundings as they become immersed in the task at hand. In November 2004, *The Times* published a short complaint in its Letters section. It read: 'Sir,

---

*In the case of the nineteenth-century physician Paul Wolfskehl, an enthralling maths problem proved to be a lifesaver. In some accounts, Wolfskehl had resolved to commit suicide after his advances towards a young lady were resisted. He set a date on which to put a pistol to his head at the stroke of midnight. That evening, while at the library, Wolfskehl stumbled upon a paper regarding Fermat's Last Theorem (which had not yet been proved). Wolfskehl became absorbed in the work, and lost the hours of the evening to contemplating the intricacies of an alleged proof. He became so engrossed, in fact, that he lost all sense of time and wandered past his self-appointed time of demise.

Sudoku puzzles should carry a warning. It's only Day 1 and already I've missed my Tube stop. Yours truly, *Ian Payn*, Brentford.' Ian Payn was not Sudoku's only victim. In June 2008, an Australian court halted a drug trial when it transpired that five of the twelve jurors had been playing Sudoku instead of listening to evidence.[40] Sudoku has since passed the early stages of a new craze and is now firmly embedded as a daily ritual the world over, drawing in casual problem solvers just as readily as committed number bods. Millions of Sudoku players find themselves enwrapped in a game that relies solely on their reasoning skills.

In the previous chapter, we saw that humans are drawn to puzzles because they fill a tantalising information gap between what we know and what we don't. But what keeps us engaged while we're solving them? Will Shortz, crossword editor for the *New York Times* and self-professed Sudoku addict, speaks of how Sudoku 'has very simple rules. You can learn it in ten seconds, and yet the logic needed to solve Sudoku is challenging.'[41] To its enchanted followers, Sudoku seems eminently solvable, but it presents a level of difficulty that makes the effort worthwhile.

The measure of a problem is in the experience and emotion that comes prior to the breakthrough moment (which is why I suggest that the circumstances of Archimedes's death say more about the man than his *Eureka* moment). The search for a solution can be long and winding, and it can afflict us with feelings of frustration, agitation and pleasure – often all at once – as we strive to make the incomplete complete. The most compelling problems leave us immersed in the hunt for a solution, a state that psychologists have put language and meaning to.

Optimal experience and flow

Have you ever been so deeply immersed in an activity that you lose all sense of space and time? You sit down to read a novel, you go on a hike, you take a friend to dinner, and before you know it, hours have passed. These are the *optimal experiences* we all strive for in our pursuit of the good life. The psychologist Mihaly Csikszentmihalyi (pronounced 'chicks-send-me-high') uses the term *flow* to signify the state of immersion in which 'people are so involved in an activity that nothing else seems to matter'.[42] Flow occurs in all walks of life – Csikszentmihalyi speaks of the feeling of a wind whipping through a sailor's hair, a painter seeing their new creation come into being and the father whose child responds to his smile for the first time (having experienced this last one not too long ago, I can confirm that it is about as optimal as life experiences get).

In the context of performance, flow speaks to how humans get the very best out of themselves. When we witness people in their element – Roger Federer striking a backhand, a marathon runner covering the ground in graceful strides, or a dance troupe that never misses a beat – we marvel at the control they have over their most intricate actions. For the subject, flow is a state of exhilaration where all the world's forces align to their commands.

Experiences are optimised, says Csikszentmihalyi, when there is 'order in consciousness'. We have already seen in this chapter that human ingenuity arises from shifting between different layers of consciousness. Flow speaks to the very pointed, very conscious effort we make to maximise our performance. Our most immersive experiences occur when we are able to shut out external cues and pour every ounce of conscious energy into the task at hand. 'Concentration is so intense', says Csikszentmihalyi, that 'there is no attention left over to think about anything irrelevant, or to worry about problems'. Csik-szentmihalyi is a very definite optimist; he believes flow is a state that we can usher ourselves towards: 'The best moments usually occur when a person's body or mind is stretched to its limits in a voluntary effort to accomplish something difficult and worthwhile.'

Flow can only be achieved when the difficulty of the task is perfectly matched to our skill. We'll more willingly immerse ourselves in a task when we perceive that it stretches us just beyond our current capabilities: a task that is challenging but attainable. In contrast, when our skill far outweighs the dif-ficulty of a task, we feel under-challenged. An excess of these tasks can only result in *boredom* – nobody delights in hammer-ing away at the same thing over and over. Executing the same task repeatedly may induce a superficial sense of mastery, but there is no joy to be had from staying in your comfort zone. We experience *anxiety* when we perceive a task to be more chal-lenging than we're able to handle: we feel over-challenged. In this case, there is nothing productive to grasp at because we lack the knowledge or skill required to bridge the gap between problem and solution.

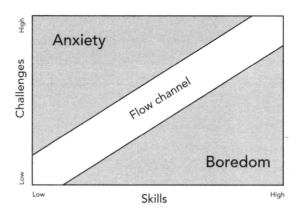

The challenge–skill framework is useful for understanding the contrasting ways in which mathematics is experienced. Because mathematical knowledge is interwoven, a small gap in one topic may amplify over time and spread in several directions, much like a Jenga tower becoming more unstable with each block that is removed. A knowledge gap in *factors and multiples*, for example, may hamper you in grasping the basics of *fractions*, a struggle that only magnifies when you later go on to study topics like *probability* that are couched in the language of fractions. Chaos theory tells us that small actions can have huge effects over time – an innocuous flapping of a butterfly's wings can cause tornadoes on the other side of the world. Chaos is what learning descends into when crucial gaps in our knowledge are left unaddressed. As small knowledge gaps explode in size, they leave us with exaggerated impressions of our struggles. In the other corner, scores of students profess outright boredom with mathematics as they work through reams of calculations, mundane and repetitive in nature (giving yet more reason to delegate to the machines: they are immune to boredom). Humans are too curious, too able, to settle for the mundane.

Students of mathematics – hobbyists and professionals alike – routinely find themselves in flow because the subject is replete with multilayered concepts and puzzles. There is joy to be had in poring over new ideas and making conceptual leaps, one increment at a time. We can feel ourselves getting wiser as we equip ourselves with new problem-solving strategies and models for thinking.

That's not to say it's an easy ride. Flow is largely a consequence of how we regulate our struggle. A problem can stump us in two ways: when we lack the relevant background knowledge, or when the solution requires a deep insight that combines knowledge in some novel way (or both, of course). To achieve flow, we need feedback loops to assess our knowledge and skill level, and to find problems that are ideally suited to both. The psychologist Anders Ericsson goes as far as citing feedback as one of the core traits of the thousands of hours of *deliberate practice* required to develop expertise: 'Deliberate practice involves feedback and modification of efforts in response to that feedback. Early in the training process much of that feedback will come from the teacher or coach, who will monitor progress, point out problems, and offer ways to address those problems.'[43]

In education, coaches go by another name: tutors. The classical model of one-to-one tutoring was established by Aristotle, who took one Alexander the Great under his wing. When the latter built a library at the Lyceum, he made sure to include many books on instruction that expounded the importance of a tutor being aware of their student's knowledge in order to expand it and to correct misconceptions. More recently, a study by Benjamin Bloom in the 1980s demonstrated that students who receive one-to-one, 'individualised' tutoring significantly outperformed students receiving conventional instruction as part of a group.[44] A tutor, in short, helps the learner to systematically acquire knowledge, the building blocks of learning. Tutors help students to identify and correct their own errors, turning each minor failure into a growth opportunity. It is notable that deep learning, the most promising area of AI research, is premised on this idea of self-correction. A deep learning algorithm is exposed to its own errors and uses this information to adjust its parameters automatically. Humans can self-correct too, but we sometimes need the guiding hand of a supportive tutor or coach to point out our mistakes and lift us to the next level of performance. A tutor prods, prompts and provokes to ensure the learner is engaged and optimally challenged at all times, selecting problems that stretch the learner to the edge of their ability and giving them plentiful opportunities to apply their recently acquired knowledge in the most creative ways.

Video games have been adopting this model for years. The most absorbing games are designed for optimal challenge. They start by assuming little background skill on the player's part, but gradually equip the player with new skills as the levels progress, exposing them to ever more ambitious challenges. We concede the night to hours of journeying through the virtual game-world because, at each moment, we are reaching for the next incremental breakthrough.

One of the hallmarks of good coaches is that they help their students to become self-sufficient. Ericsson notes: 'With time and experience students must learn to monitor themselves, spot mistakes, and adjust accordingly.' This is one way of understanding the journey from novice to expert; as we become accustomed to our craft, we rely less on external feedback loops. We learn to self-diagnose, to set our own learning paths, and to identify tasks of appropriate difficulty. We regulate our intake of knowledge. This is a disciplined act that requires smart choices about what to learn and when. At times, it obliges us to hold off on acquiring new knowledge, and instead to pause, reflect and solve with what we already know. An implicit feature of video games and tutors is that they place sensible constraints on what learners can do, and when, in order to maximise their learning.

Restraint from oracles

For a professional mathematician attacking a previously

unsolved problem, struggle is rooted in the possibility that the solution is beyond reach. Mathematician Edward Frenkel likens a maths problem to a jigsaw puzzle whose final picture is unknown; the struggle comes from the uncertainty of not knowing whether a picture will even emerge.[45] There is a rather different struggle that comes from taking on established problems for which solutions have already been discovered.

Problem solvers of the past could not have imagined enjoying instant access to all the world's information. But unfiltered access to knowledge has its drawbacks. Creative thinking emerges from constraints, and there are occasions where depriving ourselves of knowledge reaps cognitive benefits. A puzzle loses its value (not to mention its enjoyment) as a thinking exercise when the solution is delivered to us on a platter. The internet has democratised access to knowledge like no other technology. Google – the company and the product – can lay claim to much of the credit. The company was founded 'to organise the world's information' and its search engine is the closest thing today's internet users have to an oracle. In his provocatively titled article 'Is Google making us stupid?', Nicholas Carr blames the internet for his dwindling attention span.[46] He acknowledges that its supply of information is a 'godsend' but laments the fact that it is served up so instantaneously, in bite-sized morsels. This has the effect of altering our mental habits, inculcating in us a tendency to skim tasks rather than immerse ourselves fully. In our urge to grasp quick and easy punchlines, we sacrifice all-important precursors of flow like patience and reflection.

A taxing problem will leave us grappling with uncertainty, and the internet gives us the prospect of immediate resolution through a quick search. If problem solving is to serve its purpose as a conditioning exercise for the mind, we must somehow resist the urge to pluck out ready-made answers. Google is designed for frictionless knowledge transfer; it is not your coach, and it has no stock in how you experience learning. It is cold and precise in its singular aim of feeding you answers.

Admittedly, this issue predates the internet. Almost every puzzle book contains solutions at the back, designed to provide the reader with a complete, end-to-end experience. But the intent of the puzzle setter is inadvertently undermined when barriers between problem and solution are removed. To expect the reader to resist a peek at the answer adds a new variable to the mix, *willpower*, which is often in limited supply for humans.[47] It is for this reason that I usually rip out the solution pages of puzzle books – it is my way of engineering some friction, which is essential to the effort and struggle that makes for a rich problem solving experience.

Google does not just stamp its authority on serving up answers – it will even intervene as you formulate your search query. One of the most promising areas of machine learning, widely considered to be the holy grail of AI, is *natural language processing*. As technologies acquire the ability to mine through and make meaning of text, they are being deployed to complete our thoughts before they are fully typed – predictive text is already so pervasive that we would all probably admit to yielding to an automated suggestion at some point or other. This marks a shift away from tutoring behaviours – a tutor would rightly be admonished if they encroached on a student's every thought, scarcely giving them a chance to articulate an idea for themselves. Spell-checking tools that were once genuine providers of feedback, with plenty of friction built in to enable the writer to consider alternative phrasings, have been displaced by overbearing autocorrect functions.

Information technologies need to be more coach than

oracle. The supply of answers has to be tempered with a willingness to give the user feedback and options for further inquiry. Some of the best online learning content is also the most interactive, with prompts and scaffolds that encourage users to reflect on their mistakes (as opposed to spoon-feeding them the correct answers). Even calculators can become more empathetic to the learning needs of their users. The QAMA calculator (which stands for 'Quick Approximate Mental Arithmetic' and is inspired by the Hebrew word for 'how much?')[48] requires the user to provide what they think is a reasonable estimate of whatever calculation they have in mind. If the estimate is deemed 'reasonable' (the definition of which is at the core of QAMA's algorithms), then the screen shows the precise answer. If, on the other hand, the estimate falls outside of what QAMA considers a reasonable range, then it prompts the user for another estimate. The idea of withholding knowledge may appear archaic in the information age, but when done with purpose, it can force active participation in learning rather than passive consumption.

A drive from within

I never had reason to fear paperclips – until I encountered a particularly unsettling thought experiment by philosopher Nick Bostrom that speculates on how machines of the future might behave once they have surpassed human levels of intelligence.[49] Bostrom invites us to imagine a superintelligence whose primary goal is to manufacture paperclips. The objective seems innocent enough, until you start to get into the mind of such a being and unravel the unintended consequences that might follow as it creates its own sub-goals. The super-intelligence might transform all the earth's matter into gigantic paperclip-manufacturing facilities. It might even proceed to turn outer space into a supercomputer that keeps track of the number of paperclips it has produced. The thought experiment, for all its outlandishness, is a useful reminder of the singular focus that machines direct towards problems.

This chapter has sought to demonstrate that we humans, too, are capable of productively obsessing over problems. Our biological substrate gives rise to countless threats to thinking and problem solving – it is humans, not machines, who grapple with feelings of boredom and anxiety. And it is humans, not machines, who have the means to evaluate our mental dispositions. We can regulate the speed at which we think, as well as the difficulty of our tasks, to shift us into the most productive state of flow. Technology can help usher us towards flow-like states, but only when its knowledge-feeding impulses are kept at bay.

Whether or not we end up in flow, and the kind of flow we experience, is shaped by what drives us. Machines are not motivated entities: they are simply programmed to make calculated choices to minimise their errors according to mathematical models (which themselves are specified by humans). To get a computer to work, you simply push the 'on' switch. The human mind – or at least our current understanding of it – rises above these programmatic descriptions. Our 'on' switch is internal in the sense that the ideas we mull over depend very much on what commands our attention at both the conscious and the subconscious levels.

We solve problems, often by spilling blood, sweat and tears, because their resolution brings us immense satisfaction. While humans are routinely conditioned to respond to external rewards and punishments, this is not how we manifest the best versions of ourselves. Csikszentmihalyi notes that flow is more likely to be attained when we are *intrinsically* motivated by a problem, because it means we are more invested in directing our consciousness than in directing external inputs. When it comes to creative tasks for humans, intrinsic drivers are more potent motivators than external ones.[50] Intrinsic drivers make us more resilient in the face of struggle and fuel our most novel thoughts. Carrot-and-stick approaches that have proved popular among AI researchers, like reinforcement learning, may overlook a crucial insight into human intelligence – that we are more productive, more creative and more uplifted when we find purpose in the doing of tasks, and not merely in the completion of them.

7

COLLABORATION

*An unlikely mathematical duo, how ants get their
intelligence, and the quest for a super-mathematician*

To pit humans against machines is to miss the narrative arc of technology. That realisation is not born out of some whimsical hope that humans will outpace or outperform machines in calculation feats: there, the ship has well and truly sailed. Nor is it to suggest that our silicon counterparts are ready to render human intelligence moot. What we have seen instead is that machines are immensely powerful thinking partners because they possess forms of intelligence so distinct from our own. Machines augment our ways of understanding the world by offering us a particular lens through which to view it.

Part I of this book unpicked five principles of mathematical intelligence that distinguish our ways of thinking from that of machines. This attempt to separate humans from machines exposed a subtler interplay between the two: we can harness technology to amplify those facets of intelligence that we consider to be uniquely human. This is, after all, what it means to augment: it is *because* machines think differently to us that they make for such effective cognitive allies.

According to Kasparov's formulation of human–machine collaboration mentioned in the introduction, when machines and humans collaborate effectively on certain tasks the cognitive output is superior to the sum of each individual contribution.

The underlying principle behind Kasparov's formula is *complementarity*. By the very same token, the scope for human–human collaboration is vast – even more so, in fact, because we are remarkably diverse in our ways of thinking.

An unlikely duo[1]

The Cambridge mathematician G. H. Hardy was quite used to receiving letters from young pretenders claiming to have stumbled on some mathematical discovery or other. One letter, which he received in 1913, seemed no different at first. It began:

> Dear Sir, I beg to introduce myself to you as a clerk in the Accounts Department of the Port Trust Office at Madras on a salary of only £20 per annum. I am now about 23 years of age ...

The clerk went on to claim a number of 'startling' results concerning numbers. Affixed to the letter was eleven pages of mathematical scribbles that listed over 120 mathematical results, many of them vaguely worded. Some bore a faint resemblance to the kinds of theorem in Hardy's own papers, though formal proofs were lacking. Some results were surprising; others seemed, at first blush, downright absurd, such as the claim that summing all the positive integers $(1 + 2 + 3 + ...)$ results in $-1/12$. Between these outlandish claims and the understated profile of the writer, there was little to pique Hardy's interest. The letter concluded:

> Being poor, if you are convinced that there is anything of value I would like to have my theorems published ... Being inexperienced I would very highly value any advice you

give me. Requesting to be excused for the trouble I give you. I remain, Dear Sir, Yours truly, S. Ramanujan.

Srinivasa Ramanujan was born in Madras, India in 1887, when the region was still under British rule. At the age of ten, Ramanujan stood out in school for his stellar academic performance and prodigious memory skills. One teacher described the young Ramanujan's mathematical talents as 'off scale'. Ramanujan earned a scholarship to study mathematics at college, but opportunities for academic progression were sparse, leading him to take the position of an accounting clerk for the Port of Madras. In his employment, he was a human computer. On the side, however, he continued to pursue advanced mathematics.

Ramanujan took inspiration from an undergraduate textbook that he first stumbled upon as a sixteen-year-old. The text was known for its terseness – it presented progressively more complicated facts and formulae without proof. This style made a lasting impression on the precocious Indian.

Ramanujan, in turn, caught the attention of his boss, who introduced his young clerk to British expatriates. They could not determine whether Ramanujan had 'the stuff of great mathematicians' or whether 'his brains are akin to those of the calculating boy'. As they inclined towards the former, the expats reached out to mathematicians back home, though with no success. Undeterred, Ramanujan decided to write to renowned British mathematicians himself, most of whom paid him no attention. His note to Hardy was little more than a shot in the dark.

It almost missed – Hardy, too, was ready to dismiss the letter as the ramblings of an amateur. When he headed to dinner that evening, he had no conscious intent to return to

Ramanujan's notes. But something festered in Hardy's subconscious mind; he could not shake off the feeling that there might be more to the letter than met the eye. Hardy enlisted the support of his contemporary, John Littlewood. As they pored through Ramanujan's results in closer detail, they began to realise that they were in possession of some deep mathematics. Hardy would later remark that the strange formulae that filled Ramanujan's notes 'must be true because, if they were not true, no one would have the imagination to invent them'. Philosopher Bertrand Russell recalled that the next day, he 'found Hardy and Littlewood in a state of wild excitement because they believe they have found a second Newton, a Hindu clerk in Madras making 20 pounds a year'. Hardy now set his mind on bringing the mysterious clerk over to Cambridge.

After expressing enthusiasm for Ramanujan's theorems, Hardy insisted: 'before I can judge properly of the value of what you have done, it is essential that I should see proofs of some of your assertions.' Ramanujan's response was perfectly frank and honest: 'If I had given you my methods of proof,' he said, 'I am sure you will follow the London Professor [who had rejected Ramanujan's approach].' Of the claim that $1 + 2 + 3 + 4 + \ldots = -1/12$, Ramanujan said: 'If I tell you this you will at once point out to me the lunatic asylum as my goal.'

The tetchy exchange set the tone for Hardy's collaboration with Ramanujan when the latter finally arrived in London in April 1914 after a month-long voyage. Ramanujan had prepared for his trip by dressing in Western clothes and learning to eat with cutlery. It would take considerably more effort, though, for Hardy to persuade his new protégé to adapt to his own ways of doing mathematics.

By most standards, Hardy was a top-notch mathematician. After ranking fourth in the undergraduate Cambridge Tripos exams (three places below where he felt he truly belonged), Hardy devoted himself to the more formal and rigorous approach of 'pure' mathematics that was gaining popularity in Continental Europe. This is the same Hardy that gave no room to 'ugly mathematics', and for whom 'permanent' truths were the pinnacle of mathematical inquiry. Hardy's papers did not always bring forth revolutionary results, but they were exemplars in how to write mathematical arguments. Hardy took pride in fine-tuning proofs, a process that he embraced as an act of craftsmanship. He proudly adopted the view that a mathematician's role is to extend, however incrementally, the frontiers of knowledge that earlier scholars have reached. Mathematical discovery was a journey of continuous progression: it did not rely on sudden discoveries, least of all empirical ones.

In contrast, Ramanujan's spirituality (he was raised as a Hindu brahmin) had an enduring influence on his intellectual outlook. Mathematics, for Ramanujan, was a largely holistic enterprise grounded in regular leaps of faith. He delighted in manipulating equations, relying on his deep-rooted intuition and phenomenal talent for arithmetic. He would often attribute his wondrous formulae to the Hindu goddess Namagiri, who he believed had divinely revealed them, bringing them to the tip of his tongue.

There was bound to be tension between the rigorously minded Hardy and the tour-de-force purveyor of formulae that was Ramanujan. To Hardy's mind, formulae were a dubious basis for generating mathematical truths. If a formula held true, it ought to be established through generalised proof alone. Hardy was also unaccustomed to exercising leaps of faith, and certainly not ones that leaned on spirituality (this was a man who went out of his way to establish the non-existence of

God). For Hardy, the acceptance criteria of mathematics left no room for even the tiniest gaps. He would rebuke Ramanujan for playing fast and loose with infinite sums and other concepts that, to his mind, warranted strict definition.

As Hardy interrogated the masses of formulae in Ramanujan's notebooks, he presumed there must be some overarching narrative – perhaps a grand theorem – that tied them all together. He would be left frustrated when Ramanujan declared no such purpose: Hardy could not conceive that one could dream up such complex ideas without a lofty vision. For Ramanujan's part, the need to justify his every claim seemed strange. Were the Europeans so insecure in their arguments that they had to tediously check every step?

With time, the two men embraced one another's style. Hardy even made a conscious effort not to force formal instruction on Ramanujan, realising that this would only stifle the young genius. This compromise worked wonders: the two men ultimately bridged their intellectual divide to the extent that Hardy would later describe their collaboration as the 'most romantic affair' of his life. In his twilight years, Hardy would even espouse the virtues of thinking holistically, a sentiment he almost certainly borrowed from his Indian counterpart. Hardy seemed to accept that the purpose behind Ramanujan's formulae was something the young maestro could only intuit, and that he struggled to express them in formal terms.

Ramanujan's stay in Cambridge was cut short as the world confronted the horrors of the First World War, during which he contracted tuberculosis. It would spell the beginning of Ramanujan's end, and he passed away soon after his return to India in 1920. The Ramanujan–Hardy collaboration lasted six years, starting with Ramanujan's letter. Many of their findings continue to fuel inquiries in number theory, and some have even been found to have practical value long after their deaths (Hardy might not have been so thrilled by this: he proudly attested to the uselessness of his results). Modern-day programs like Wolfram Alpha, for example, make explicit use of Ramanujan's formulae to compute the digits of π.

Hardy and Ramanujan exemplify what is possible when two humans collaborate with one another. In particular, two humans who think in complementary ways and bring wildly different perspectives may combine their talents to great effect. It is natural to ask what happens when *n is greater than two*: in what manner might collaborative potential increase as we train more human minds on a problem?

Emergence: when the whole is greater than the sum of its parts

The longstanding notion of intelligence is centred on the individual, typified by nineteenth-century historian Thomas Carlyle's 'Great Man'.[2] Yet celebrations of lone wolves often mask the collective contributions of a band of people lurking in the background. Michelangelo receives sole credit for his masterpiece that illuminates the ceiling of the Sistine Chapel, despite the fact that he rallied thirteen people to work on the painting under his watchful eye, along with more than two hundred assistants when working on the Laurentian Library in Florence. Historian William E. Wallace aptly terms the maestro a *CEO*, his works a triumph of coordinated entrepreneurship.[3] Thomas Edison was no more alone in his inventive feats, though he would often go to great lengths to wrest the plaudits from his brilliant employees. The most non-trivial problems require the weaponry of groups rather than the wit of lone rangers.[4] In recent years, the idea of a group's 'collective intelligence' has been gaining steam. To predict the performance of a group on a task, you are often better off evaluating their combined intelligence rather than totting up their individual smarts.[5] But what distinguishes these groups from ordinary collections of people? To answer that, we turn to the insect world.

By any reasonable standard, an ant is stupid. It has the paltriest of brains (250,000 cells versus 86 billion for us humans), which means it has no serious ability to think, reflect or plan. Despite this, large groups of ants combine into colonies exhibiting behaviours that are unequivocally smart. As colonies, ants are capable of remarkable feats. They can find food and reproduce themselves. They can maintain farms of fungi and take care of 'cattle'. They even wage war and defend themselves. How is this possible, given the uncontested stupidity of each ant?

Colonies operate according to very simple rules. Let's take the example of how an ant colony goes about distributing jobs.[6] Suppose the colony has an even split of workers, caretakers, soldiers and gatherers – each comprising a quarter of the total colony. When two ants meet, they are able to identify one another's jobs by using their antennae to pick up scent – different smells for different jobs. By recognising patterns in pheromone trails, an ant is also able to keep track of the rate at which it has met with ants with other jobs, and it uses this information to determine its own task. Imagine, for instance, that an anteater comes along and kills all the gatherers, upsetting the balance of the colony. A worker ant will continue to meet other ants, except now those ants will be either caretakers or soldiers. With each interaction, the worker ant will realise that there is a shortfall of gatherers, and eventually it will take up that role. The balance of jobs in the colony thus restores itself. The ants are not getting their orders from a supreme queen ant up top (the queen may be fed and cared for by neighbouring ants but has no way of physically communicating with those further away). Rather, they are aggregating information across a vast network of interactions that enables the colony to be productive.

Ants are among a group known as social insects (whose members also include bees, wasps and termites). They are so named because they thrive in large numbers – collectively, social insects form more than half of the earth's insect biomass. Not so stupid after all.

Ant colonies are an example of *emergent* behaviour. Emergence refers to 'the arising of novel and coherent structures, patterns and properties during the process of self-organisation in complex systems'.[7] It explains a wide range of phenomena[8] beyond ant colonies, such as how prices are set in the marketplace by aggregating the behaviours of consumers and suppliers, how individual molecules of water combine to give the property of wetness, and how each of those 86 billion neurons in the human brain are capable, as a collective, of producing complex thoughts and memories, even consciousness. And even while some of these behaviours have yet to emerge in machine learning programs, they too rely on the same idea of building intelligence from the bottom up, combining the simplest elements to produce higher-level behaviours.

Emergence is not simply a device for elevating the stupid. Intelligent beings – humans, for instance – can apply the same principles to attack complex problems through networks. The first requirement is numbers: when it comes to problem solving, no person is an island. People tend to overestimate their knowledge of everyday things – what psychologists term the 'illusion of explanatory depth'.[9] The classic example is a zip fastener: when people are asked whether they can explain how one works, they usually profess to have more knowledge than they can actually demonstrate when prompted to give an explanation. It is in our human nature – an artefact of our cognitive miserliness – that we gloss over the rich details of how most things work. We generally think we know more than we actually do about the world. It is only when we are put to task, as in the zip example, that we realise how much knowledge we store in our surroundings. There is a mismatch between the complexity of our problems and our brains' modest storage capacity, which means we have to rely on our bodies, the environment and other people to access the knowledge we need. Working with other people is simply a way of dividing the cognitive labour required to carry out mental tasks.

The idea of groups outperforming individuals has a well-documented history.[10] At a country fair in 1907, 787 members of the public entered a competition to guess the weight of an ox on display (more recent renditions of the experiment involve estimating the number of jelly beans in a jar – the conclusions are the same). The statistician Francis Galton turned the competition into an impromptu experiment by analysing the fairgoers' guesses. When he looked at the distribution of guesses, Galton found that they lay on a bell curve (most estimates were somewhere in the middle and just a few at the lower and upper extremes) – much in line with what he'd expected. What surprised him was that the average estimate was almost bang on the money – the subjects had ventured 1,197 lb on average,* just one pound short of the ox's true weight. 'The result seems more creditable', concluded Galton, 'to the trustworthiness of democratic judgement than might have been expected.' In other words, just like ants, groups of people can

---

*Galton adopted the median as his measure of central tendency to avoid distortions due to outlier estimates.

exhibit behaviours more intelligent than the sum total of the individual intelligences among them.

But numbers alone are not enough to foster productive collaboration. Emergent behaviour is no free lunch: it does not arise simply by grouping together lots of elements. The African elephant has three times as many neurons as humans, but it does not speak, write poetry or formulate mathematical proofs. Of much greater significance are the structures that connect those elements; our intelligence is a function of architecture as much as neuron counts.[11] Poorly designed networks can even lead to disastrous outcomes. Naturalists have found instances of huge armies of ants that simply move around in large circles repeatedly until they drop dead. Biologists call this a 'circular mill', and it arises when ants become separated from their colonies, at which point they obey the simplest rule: *follow the ant in front.* The circular mill breaks only when a cluster of ants somehow strays, and others follow. When the mindless, unproductive rule is carried out by every ant, the aggregate result is inadvertent mass self-destruction. Group behaviours can be channelled in directions both positive and negative: they can distribute jobs within ant colonies, or lead the same ants on a steady march towards certain death.

Just like ants, humans can also influence one another to our collective detriment. In another experiment,[12] subjects were shown an unnumbered line and asked to indicate which numbered line matched it. When the subjects worked alone, they had high rates of success. In other cases, five people entered the room, all of whom were actors (unbeknownst to the subject) and gave the same wrong answer. Those subjects, after some hesitation, had much lower rates of success, with as many as one third succumbing to the stooges' incorrect answer. This is another instance of what we learned about the causes of human bias in Chapter 3: our reasoning is often shaped by persuasion dynamics, because our survival depends on staying in the herd. The term 'groupthink' was coined by psychologist Irving Janis to describe the social phenomenon in which 'members of any small cohesive group tend to maintain esprit de corps by unconsciously developing a number of shared illusions and related norms that interfere with critical thinking and reality testing.'[13] Collaboration is fuelled by reasoning about other people's mental states, and experience shows that we will gladly shy away from questioning assumptions, lining up behind flawed arguments on the basis that they represent the majority view – the human equivalent of the ants' death march.

Why diversity matters

How do we make sure that the double-edged sword of human–human collaboration tilts in our favour? *Diversity of opinion* is critical – it occurs when 'each person [has] some private information, even if it's just an eccentric interpretation of the known facts'.[14] When this happens, each person's independent judgement combines, to great effect. Their errors cancel each other out. Those fairgoers each contributed a morsel of information based on their own unique life experiences. The butcher might recall the last ox he handled. The avid ox enthusiast may happen to have read about the typical weight of an ox. The everyday meat consumer would just go on the bite-sized portions they swallow up. The veggie lover may have no benchmark at all and will appeal to intuition. No guess is perfect because no life experience is complete, but if the group is diverse enough then the collective knowledge pool is so vast that individual errors tend to nullify one another, resulting in a remarkably accurate overall average estimate.

If we are relying on other people to shatter our illusions of explanatory depth, then we must make sure that their knowledge is not simply a mimicry of our own. It must extend our worldview rather than amplify it. Where homogeneous groups of like-minded thinkers tend to exploit the narrow set of ideas they already share, heterogeneous groups are able to combine their different perspectives to expand their mental horizons. This holds even at the molecular level. Our digestive system alone relies on several different proteins, each dedicated to a major food group – amylase for starch, lipase for fats and so on. No single protein can break down everything; we rely on their collective capabilities.[15]

This 'cognitive diversity' is a prized asset of collaborative groups,[16] and never more so than with interdisciplinary problems that demand multiple perspectives. From an evolutionary perspective, cognitive diversity is essential to a population's survival: every society needs a mix of adventurers pioneering us towards new discoveries. We also need risk-averse people, and a whole spectrum of dispositions in between, to strike the right balance between exploring new frontiers and exploiting the resources already at our disposal.

The more cognitively diverse a group – the more varied its perspectives and ways of processing knowledge – the more its collective intelligence outstrips the sum of its individual parts (a particular instance of this is that the proportion of women also predicts better group performance).[17]

Take the Covid-19 pandemic, effective responses to which drew on experts from wide-ranging areas, including public health, epidemiology, virology, immunology, primary care, intensive care, behavioural science and economic policy. Mathematical modellers were part of the mix too, projecting a range of scenarios as new evidence of the virus's potency poured in. In the UK, mathematicians were very much in vogue: just prior to the pandemic, the Prime Minister's most senior adviser, Dominic Cummings, posted a blog declaring that he was looking to recruit 'weirdos and misfits with odd skills' in an attempt to apply more scientific thinking to the civil service. Cummings was thinking of mathematically minded data scientists, so their presence on the government's Scientific Advisory Group for Emergencies (SAGE) was assured. Good news for weirdos, you might think, except that the newfound fervour for mathematicians (most of whom do not consider themselves weirdos) seemed to come at the expense of more pluralistic approaches in dealing with Covid-19. In an article published in *Nature*, twenty-two signatories gave a stinging rebuke concerning the politicisation of mathematical models to justify questionable policies, suggesting that too much faith was being placed in the precision of data-driven projections.[18] The paper makes clear that mathematical modelling has its place as one of several connected disciplines, but that it should not rule over all domains for an issue as complex and multifaceted as a global pandemic. It transpired that SAGE had no virologists, immunologists or intensive care experts on its team (it also had just seven women among its twenty-three members), sparking concerns that the government's response mechanisms were limited to a handful of select perspectives.[19] Alternative response groups formed to address the cognitive diversity gap.[20]

Covid-19 models themselves exploit the power of diversity. Epidemiologists frequently make use of *ensemble* forecasts which, as the name suggests, forge predictions by combining multiple models. It is another wisdom-of-the-crowd mechanism, where each model gets to have its say and the final prediction is determined by a voting mechanism. Ensemble models tend to outperform their constituent parts when the individual models have a degree of volatility; the ensemble somehow captures the best elements of each model while washing away their more erratic behaviours. Each individual Covid-19 model derives from the data and assumptions of each modeller. It's not a question of whether mathematicians (or weirdos) outperform others; it's a question of how to bring diverse groups of modellers together to produce forecasts that outperform those of any single group.[21]

Ensemble models are also popular in AI, where groups of algorithms are meshed together to outperform individual ones. In a similar vein, we've seen that old-fashioned symbolic AI, with rules hard-coded, is being hybridised with modern-day machine learning algorithms – two vastly different approaches to automating intelligence, the combination of which is increasingly being heralded as superior to either one alone (as well as more reflective of human intelligence).

Cognitive diversity derives its power from the enormous range of representations that each model brings to bear on a situation. That leaves humans, with our incredibly rich and diverse sociocultural heritage, poised to flourish when we join forces to solve problems. We saw in Chapter 2 how mathematical constructs such as our counting systems are interwoven with our experience and environmental backdrop. Experiments carried out by psychologist Richard Nisbett go even further, demonstrating how culture profoundly influences the way we see the world.[22] Westerners and East Asians, for instance, seem to direct their attention in different ways. When shown vignettes of various scenes (among them a train, a tiger in a forest and an aeroplane surrounded by mountains) Americans tend to home in on the focal object, whereas Japanese subjects are more likely to map out the whole scene, fixating just as much on background detail. Other studies have similarly teased out the interplay between culture and information processing by examining how subjects respond to optical illusions.

Consider the two middle circles below.[23] Participants from industrialised nations are more likely to (incorrectly) say that the right-hand circle is larger; in fact, they're the same size. If you remove the outer circles, this illusion disappears. The error comes from considering the relative size of the inner and outer circles. Participants from more 'traditional' societies have less grounding in abstraction and are therefore less likely to be influenced by the relationship of the middle circles to the outer circles. They have much higher success rates with this question, although the trend is reversed on problems that rely on abstraction (such as those popular in IQ tests).[24]

The point here is not that one way of seeing things is superior to the other, but rather that our complementary viewpoints expand our collective understanding of a problem. If ever there was a case for multiculturalism, it is surely here: by exposing ourselves to different ways of living and being, we acquire a richer tapestry of mental models and free ourselves from monolithic thought patterns.

It's a troubling irony that for all the technical effort to diversify AI models, the field itself remains narrowly represented. Over 80 per cent of machine learning practitioners identify as men,[25] while black workers make up less than 5 per cent of staff at Google, Facebook and Microsoft.[26] There is a traceable path between the narrow demographics of AI developers and the biases inherent in their technologies, yet Big Tech firms have so far paid mere lip service to such concerns. In one high-profile incident, Google's Timnit Gebru was forced to resign her position following an internal review of a paper she had co-authored that highlighted the discriminatory nature of natural language models employed by the company's search engine.[27] The silencing of minority voices is a licence for AI to project and amplify our latent human prejudices onto the world as the assumptions of a small band of innovators are left unchecked.

Think back to the unlikely partnership of Ramanujan and Hardy: one a Hindu Brahmin, the other an ardent atheist; one who became infatuated with exotic formulae from a textbook, the other inspired by the European paradigm of rigour; one thinking holistically, the other demanding formal proof. Both mathematicians brought their respective representations of the world – underpinned by their upbringing and environment, and every facet of their educational experience – to bear on the same problems. Because they heralded from such disparate backgrounds, they were able to enrich one another's perspectives. The unlikeliness of the collaboration is precisely what made it so potent.

The arc of science bends towards collaboration

Science is a collegiate affair that requires multiple roles and perspectives. Contrary to popular depictions, it is not the sole preserve of lone geniuses, working in isolation in underground labs in search of the next pioneering breakthrough. A community of peers is needed to define, collectively, what problems are worth solving in the first place. As solutions are presented, review committees must convene to determine whether a submitted paper is accurate and insightful enough to merit publication. Even the most individualistic pursuit of, say, a Nobel Prize, or its mathematics equivalent, the Fields Medal, is contingent on recognition that is bestowed by panels of expert judges. No modern-day scientist earns their fame in a silo.

Every scientist 'stands on the shoulders of giants', as Isaac Newton famously put it, iterating on the foundations laid down by prior generations. These days, most worthwhile problems are interdisciplinary in nature, requiring teams of scientists who pool their diverse solution strategies. Studies dating back to the 1960s show that the most prolific scientists, and the most celebrated (Nobel laureates, for example), are also the most collaborative.[28] Social scientist Etienne Wenger puts it best: 'Today's complex problem solving requires multiple perspectives. The days of Leonardo da Vinci are over.'[29]

There is plenty of empirical evidence to suggest that science has become increasingly collaborative in recent times. An influential 2007 study by Kellogg School professors Brian Uzzi and Benjamin Jones, in which the authors analysed almost 20 million research papers in the Web of Science database, noted a 'shift towards teams' since the 1950s, and that 'teams were not only becoming more prominent, but they were becoming bigger each year ... teams were also, across a majority of disciplines, increasingly producing the most impactful papers'.[30] An analysis of the PubMed database of papers in the biomedical and life science fields has demonstrated a five-fold increase in the number of authors per paper between 1913 and 2013, and has projected that by 2034 papers will have an average of eight authors.[31] The study acknowledges the rise of so-called 'big science' projects like the Large Hadron Collider and the Human Genome Project. Taking an example from the former, it is somewhat ironic that the *Higgs boson* is named after the individual (Peter Higgs) who first postulated its existence, when the two papers that *confirmed* its existence comprised over 5,000 authors representing dozens of institutions and countries (the papers came in at around thirty pages each, of which around nineteen pages were the author list alone).[32] While Peter Higgs may have deservedly scooped the Nobel Prize for his contribution, recognition has also been distributed to the thousands of engineers, theorists and lab technicians whose individual roles combined to make the discovery possible. The term *hyperauthorship* has even been coined to describe the phenomenon of mass collaboration that arises when problems demand expertise from so many minds.[33]

Mathematics has followed the same trend of dramatic increases in collaboration over the past century. Between the 1940s and the 1990s, the proportion of authors involved in joint papers climbed from 28 per cent to 81 per cent.[34] The average number of collaborators of an individual author also rose from 0.49 to 2.84. One of the biggest advocates for collaboration in mathematics was twentieth-century Hungarian mathematician Paul Erdös. Erdös had the utmost respect for maths problems (and for coffee, embodying the quip that mathematicians are devices for turning it into theorems). Recognising that problem solving is not an individualistic pursuit, Erdös actively sought out fellow problem solvers as he travelled the world living out of a suitcase, notching up over 500 collaborators en route (many of his co-authored papers continue to be published posthumously). In fact, Erdös was so prolific as a collaborator that mathematicians identify with an *Erdös number*, which is a measure of one's 'collaborative distance' from the Hungarian – so those who co-authored with him have an Erdös number of 1, those who co-authored with someone who co-authored with him have an Erdös number of 2, and so on. It is the mathematician's equivalent of Hollywood actors' degrees of separation from Kevin Bacon.*

The mathematician William Thurston went as far as to tie the very purpose of mathematical study to collaboration: 'In short, mathematics only exists in a living community of mathematicians that spreads understanding and breathes life into ideas both old and new. The real satisfaction from mathematics is in learning from others and sharing with others. All of us have clear understanding of a few things and murky concepts of many more.'[35]

Mathematical intelligence is tied to our social constructs. We reason to persuade others of our beliefs. We create models of how other people think and behave, and we construct knowledge representations to communicate difficult ideas. We ask and answer questions that we think others will find interesting.

It is a rare mathematician who locks themselves away in isolated pursuit of a solution. Andrew Wiles may qualify as that rare specimen; notoriously, he chipped away at the proof of Fermat's Last Theorem in secret over a period of seven years. But even Wiles realised he had to enter the public fold sooner or later. He delivered a series of lectures as a way of announcing the proof and opening it up to scrutiny among his peers (which exposed a gap in his argument that would take another year to address). For all his individual brilliance, Wiles relied on stitching together results that had accumulated over the 358-year search of the proof. His was the definitive breakthrough, but it stood on the shoulders of the mathematical giants who came before. The mathematician Ken Ribet, who himself made

---

*Bacon has appeared in such a variety of films that a parlour game asks you to link any given actor or actress to him in terms of shared movie appearances – the standard challenge is to make the link in six moves.

strides that Wiles's proof leaned on, noted how odd he found it that Wiles would work in such clandestine fashion:

> This is probably the only case I know where someone worked for such a long time without divulging what he was doing, without talking about the progress he was making. In our community people have always shared their ideas … Mathematicians come together at conferences, they visit each other to give seminars, they send e-mail to each other, they ask for insights, they ask for feedback. When you talk to other people you get a pat on the back, people tell you that what you've done is important. It's sort of nourishing, and if you cut yourself off from this then you are doing something that's probably psychologically very odd.[36]

The complexity of modern-day mathematics, much as with big science, means that Wiles's example will remain an outlier among problem solvers. It is Erdös's collaborative spirit that really caught the mood of twentieth-century mathematics, with mathematicians pooling their specialised knowledge from different fields to tackle the subject's most stubborn problems.

One of the most extraordinary mathematical results of the twentieth century was also one the most collaborative. It concerned *finite simple groups,* which are the building blocks of abstract algebraic structures. The theorem sought to classify all such groups, and this classification was seen as the holy grail in the field. Just as the chemist studies molecules through atoms, and the number theorist studies whole numbers by probing the peculiar properties of prime numbers, the algebraist looks to finite simple groups as the most fundamental object of study.

The collaboration was instigated via a series of seminars

in Chicago in 1972 that outlined a vision for weaving together multiple strands of mathematics that would cover all the possible ways in which finite simple groups arise. What made the proof so remarkable was its size and form: it sprawled over 10,000 pages of papers that were sprinkled across 500 journal articles, with 100 authors from around the world. Mathematical proof had never been so unwieldy. The sceptics could not fathom that the proof would be free of error, and after much scrutiny some mistakes were indeed found – and duly corrected in the ensuing years. One of the main contributors wrote in 2004: 'to my knowledge the main theorem [of our paper] closes the last gap in the original proof, so (for the moment) the classification theorem can be regarded as a theorem.'[37] There are mathematicians today working to simplify the proof.

A telling feature of the proof is that no single person understands it in its entirety. The absence of a single, omniscient authority speaks to another precondition for emergent behaviour: *decentralisation*. The chain of command within an ant colony does not consist of a single queen ant transmitting orders to her army of subjects. Instead, it is distributed throughout the colony as individual ants literally sniff their way to localised actions. The collaborations of big science and mathematics similarly empower people to draw on their local pool of knowledge to contribute towards global outcomes. The same model of collaboration is transforming the modern-day work environment. The AI pioneer Norbert Wiener considered organisations to be 'flesh and blood machines' and believed that the intelligent potential of humans is squandered when we are confined to fixed roles:

... if the human being is condemned and restricted to perform the same functions over and over again, he will not even be a good ant, not to mention a good human being. Those who would organise us according to permanent individual functions and permanent individual restrictions condemn the human race to move at much less than half-steam.[38]

Wiener's advice is alive and well in modern companies that are tearing down rigid hierarchies in favour of more fluid structures that promote overlapping activities between departments and cross-disciplinary thinking.[39] A common feature of today's innovative companies is that teams operate as small units, with the autonomy to set their own objectives and working practices. Each unit carves out its own unique subculture, reducing the risk of excessive conformity – and the blind spots that come with it – across the entire organisation. Units can team up on demand, combining their skills and perspectives to tackle multidisciplinary projects. Even senior executives, it seems, are gunning for the virtues of emergence.

Make way for the crowdsourced super-mathematician

At the close of the twentieth century, the precedent of collaboration was well established in mathematics. The complexity of research demanded that mathematicians break through traditional barriers to collaboration and nurture their collective intelligence. The advent of the internet at around this time slotted neatly into this context, as the technologies of the digital age promised to connect people like never before. Exploiting *diverse opinions* within *decentralised structures* was a challenge the internet was made for.

Web technologies have given rise to an explosion of information as well as the tools for us to connect – both with the information itself and with those who create it. Oracle-like engines such as Google are, for the most part, pointing to knowledge produced by us humans (even as digital content becomes increasingly automated, the most interesting questions – and answers – are left to us for now). Through social media, billions worldwide are entering into a shared conversation, taking place in real time, on all topics imaginable. Economists Andrew McAfee and Erik Brynjolfsson use the term *emergence of the crowd* to refer to 'the startlingly large amount of human knowledge, expertise, and enthusiasm distributed all over the world and now available, and able to be focused, online'.[40] They recall the failed attempts in the early days of the internet to tightly regulate the production of online content. Sites like Yahoo! were soon overwhelmed by the sheer volume of human-generated content, whose growth was exponential. They liken today's web to 'a crowd-generated library – a huge, sprawling, constantly growing, and changing one'.

A messy internet also plays host to some of the most effectively coordinated projects in human history, however. Open-source initiatives have mass collaboration at their core. Source code and product blueprints are shared with the public, who are then free to modify the original design and publish their own version, releasing it back to the community. Operating systems (e.g. Linux and Android), browsers (e.g. Chrome and Firefox) and database management systems (e.g. MySQL and MongoDB) are just some of the high-profile projects that have developed in this way.

Open-source projects have also helped to bring some order to humanity's collective knowledge. Forums like Reddit, Quora and Stack Exchange and repositories like Wikipedia (dubbed the 'last best place on the internet'[41]) are exemplars of web-enabled emergence: simple agents (the lay contributor) working with simple rules (governance protocols) to collectively produce work beyond their individual capabilities. All rely on the spirit of volunteerism; an intrinsic and altruistic drive that enables the mass proliferation of high-quality, reliable content.

These examples demonstrate how within decentralised structures crowds can be trusted to create and maintain the standards of certified experts. In fact, crowds can go further. The illusion of explanatory depth afflicts us all, so convening a small handful of experts inevitably gives rise to blind spots; when the number of collaborators is stifled, so too is their collective sense of what they know and do not know. The opt-in crowd brings volume and a diversity of perspective that is unmatched by any chosen few.

The same model is being exploited in some quarters to advance the frontiers of mathematical research. In this case, the agents may not be so 'simple' – anyone who is able to contribute to an unsolved maths problem must possess serious pedigree. Regardless, in his 2009 blog 'Is massively collaborative mathematics possible?',[42] mathematician Tim Gowers wondered if large-scale problem solving could be made the norm with the tools of online collaboration. The internet diffuses information virtually in real time, which means that, at the very least, mathematicians are becoming increasingly aware of one another's work. But what if there was a deliberate effort – a forum of some kind – to bring more minds to bear on a given problem? Gowers took inspiration from the classification of finite simple groups and wondered how the approach might be updated to exploit the tools of the digital age. As he suggested: 'The idea would be that anybody who had anything whatsoever to say about the problem could chip in.' Like Wikipedia, this effort would lean on volunteers and the intrinsic drive to share knowledge in the pursuit of new horizons. It would exploit the diversity of skills and perspectives that you don't always get within the traditional confines of academia, allowing more people – and more types of people – to weigh in:

> Different people have different characteristics when it comes to research. Some like to throw out ideas, others to criticise them, others to work out details, others to re-explain ideas in a different language, others to formulate different but related problems, others to step back from a big muddle of ideas and fashion some more coherent picture out of them, and so on. A hugely collaborative project would make it possible for people to specialise … In short, if a large group of mathematicians could connect their brains efficiently, they could perhaps solve problems very efficiently as well.

It was also Gowers who wrote of the 'two cultures' in mathematics – problem solvers and theory builders. Recall, too, Freeman Dyson's characterisation of mathematicians as either visionary birds (seeking to tie together concepts) and frogs (single-mindedly focused on one problem at a time). Each occupies different vantage points, wields different tools and possesses different perspectives. Gowers was seeking a harmonisation of mathematical subcultures and felt that the internet afforded him the tools to do it.

Gowers had the digital media of blogs, wikis and forums in mind – the medium was not only to facilitate but also to shape collaborative problem solving. He meticulously laid out the rules he felt would maximise virtual collaboration. Comments would be kept short and easy to read. Etiquette would be followed – all ideas would be welcomed, even the stupid ones (and the term 'stupid' was never to be used when referring to another's work). A collective pseudonym would be used to author any published papers that came out of the collaboration, with a link to every comment. Of particular note is Rule Six, which is an ode to the principles of emergence: 'The ideal outcome would be a solution of the problem with no single individual having to think all that hard. The hard thought would be done by a sort of super-mathematician whose brain is distributed amongst bits of the brains of lots of interlinked people.' Gowers struck a chord with the mathematical community; his post earned over 200 comments, with world-class mathematicians like Terence Tao subscribing to the concept. The *Polymath Project* was born, and in his follow-up post, Gowers took aim at a particular problem.*

So, did it work? Has the super-mathematician emerged from virtual collaboration premised on simple tools and even simpler rules? There is reason to be optimistic. Within seven weeks, Gowers declared that his first problem was 'probably solved',[43] and all doubt was removed within three months. Over forty people had contributed, to varying degrees, to the solution. The Polymath Project has also given rise to published papers that marked significant progress towards two other unsolved problems. The super-mathematician appears to be emerging, one small proof at a time. Similar initiatives targeted at high-school and college students have followed suit,[44] each leveraging the powerfully simple notion that mathematicians are most potent when they solve together.

---

*In fact, the problem – known as the Hales–Jewett theorem – had already been solved. Gowers was after an alternative proof that used a combinatorial approach.

Shared intentions

There is room for all of us at the rendezvous of problem solving. The non-routine problems of the future will demand knowledge and skill that is distributed among a multitude of intelligent entities. MIT's Joichi Ito has proposed a notion of 'extended intelligence' that views intelligence in terms of an ever-evolving, ever-expanding network.[45] AI itself contributes several bits and pieces, as a vast range of models are developed and combined in order to attack novel problems. But humans too, and humans especially, enrich that network through our countless ways of seeing the world, each shaped by our individual experiences, and the distinct representations we carry with us. You and I are both part of that network, along with every other person on the planet.

The most potent problem solvers among us will be the most collaborative ones. They will possess the metacognitive awareness to know what they know and, more crucially, what they remain ignorant about. Problem solving will become an increasingly social activity as we seek out minds that complement our own. We will partner with computers in all the ways described in previous chapters. But we will also use the connective technologies of the digital age to forge closer intellectual ties with our own kin. Every human mind is a unique mould of language, environment and experience. It sometimes takes an unassuming Indian to remind us of our untouched mental capacities – one can only wonder how many Ramanujans are lurking in our midst, waiting to unleash themselves on the problems that matter to us.

No individual can bear the sole responsibility of keeping humans relevant in the machine age. Our human strengths, even our values, emerge as a collective. Scientific inquiry will be kept bright by our shared intentionality, one of the traits that distinguishes humans in the animal kingdom. Experiments show that 18–24-month-old children and chimpanzees have comparable rates of success on a range of tasks, but when the tasks are altered slightly to require collaboration with the experimenter, the success rate of children shoots up, while that of chimps dwindles.[46]

Humans ascribe value to problems based on their importance to our community. Would a ten-year-old Andrew Wiles have become so captivated by Fermat's Last Theorem had it not eluded mathematicians for so long and earned such infamy within the field? Would any scientist today be driven to work on problems that had no human precedent, or brought no recognition? Our motivation to solve problems is itself an emergent phenomenon, derived from our shared experiences. What matters to us is predominantly a function of what matters to others.

Technology has its roles to play as both collaborator and connector. We will solve with machines at the same time that the internet amplifies human–human collaboration. These technologies, in turn, will be shaped by how we humans come together; algorithmic bias is often just a reflection of homogeneous development teams. Our faith in automated judgements will be well placed only when the creators of these technologies are as diverse as humanity itself.

What technology will not do is share in our intentions. The machines are no more a part of our community than public transport is (self-driving or otherwise). They can serve our goals, but it is for humans, alongside other humans, to truly shape them. We should find comfort in the thought that however complicated the world's problems become, none of us will be expected to solve them alone.

EPILOGUE

Let's address the elephant in the room.

What if AI achieves the wildest ambitions of its creators, and machines develop human-like, or superhuman-like abilities? What if they acquire the principles of mathematical intelligence? What will be left for humans?

I have not, after all, argued that those principles are beyond the scope of what machines may one day prove capable of – only that they have so far failed to demonstrate them. I have taken particular aim at approaches such as machine learning because they are the vanguard of present-day AI and, while they result in spectacular standalone feats, they remain grounded in mindless calculation and pattern recognition that precludes comparison with our own organic forms of intelligence. I do not believe mathematics will face the 'bitter lesson' that its thorniest problems will succumb to searching for patterns in large troves of data.[1] The questions and answers of mathematics are too vast in scope, too deep in inquiry, for today's smartest machines to take the helm.

But machine learning isn't the only game in town. AI is replete with sub-fields and approaches to cracking the harder questions of intelligence – questions of how to imbue machines with common sense, reasoning, explainability, curiosity and more. It's only appropriate to countenance the possibility that much of this will come to pass.

If AI does subsume our richest mathematical thinking skills, we may take solace in the irony of it all. From the outset, computers have been conceived and developed as mathematical objects – toolboxes for programming and logically manipulating our thoughts. AI itself is just a specialised product of mathematical thinking. If we get to the point where our digital creations supersede our own thinking skills, we could consider it a mathematical job well done. Except it may be the last job mathematicians are paid for.

Unpaid work

If AI gets anywhere close to its aims, we'll see dramatic shifts towards an automated workforce. It is unlikely that humans would have the skill to spar with our machine counterparts. How presumptuous must we be to think that the machines will require, let alone accept, our commands anyway? It seems more likely that the days of the mathematician-worker will, if you'll forgive the pun, be numbered.

Then again, gainful employment was never the express aim of mathematics – an enabler, sure, but not its raison d'être. That we would get paid to think in the most creative and uplifting ways was only ever a welcome by-product, one facet of the subject's unreasonable effectiveness. Mathematical intelligence is so deeply woven into our ways of thinking and being that as long as there is a place for humans in this world, there is also a place for us to do mathematics.

That the computers may one day outperform us in certain tasks is no reason to stop engaging in them altogether. Our awareness of our own human limitations is our inspiration to play at the boundary of what is thought possible. The two-hour marathon barrier is seen by many as the next major milestone for endurance athletes.* By human standards, it is an unfathomable target.[2] By the standards of technology, it is pitiful – an amateur on a rickety bike would cover 26.2 miles in less time. Elite distance runners are not competing with an absolute standard alongside technology as much as they are striving to inch forward the relative standard of humans. It must also be noted that their efforts are hardly divorced from technology – from running shoes to data-driven training regimens, modern science is playing its hand in this most human of endeavours.

The work that humans value most is that which inspires us to create the best versions of ourselves. The ancient Greeks termed it *eudaimonia*, the 'good composed of all goods; an ability which suffices for living well'. This is a life philosophy predicated on human flourishing. A subset among us may receive remuneration for our relative talents, because of the enrichment and entertainment it offers to our fellow humans. But even if we are not paid world-class performers, the intrinsic value we attach to these activities will spur us on, just like the millions of amateur marathon runners who take to the streets each year, each pursuing their individual targets.

An automated workforce will occasion a shift in emphasis from the utility of mathematics towards an appreciation of mathematics as one of the *classics* – a subject still worthy of our study because it conditions our minds and celebrates our rich heritage as a species of thinkers, even if the job prospects are uncertain.

As people seek to carve out non-economic identities outside the workforce, we'll gravitate to activities that we ascribe

---

*The Kenyan runner Eliud Kipchoge achieved the milestone in 2019, although the highly choreographed attempt (including rotating pace setters) meant it was not recognised under official IAAF rules.

inherent value to. The future, just like many instances in the past, may see the resurgence of a leisure class, who, through the fruits of technology, find themselves with the time and space to engage in recreational thinking. Mathematics – the brand espoused by this book rather than the cold, unsympathetic variant predicated on calculation – can supply endless hours of enrichment for the mind.

The recreational brand of mathematics will even be an outgrowth from technology. We will continue to use technology to create new puzzles and to disseminate them to the world – what is the value of a puzzle, after all, if there are no takers? There will always be value in shared human experience, and technology is the means of forging bonds between tribes of problem solvers.

As we gaze up at the towering intellectual feats of machines, meanwhile, the more talented among us (who may once have found a calling as professional mathematicians) will take on the challenge of grasping their key insights. Machines may produce mathematics of an unprecedented type – more complicated, more abstract, probably less comprehensible to humans. We may face the tantalising proposition of knowing that juggernauts like the Riemann hypothesis have passed from conjecture to proof in the minds of machines, only for the details to elude our understanding. If a theorem falls down and no human has the language to grasp its proof, does it still count as a theorem? Philosophers of mathematics will never tire of such debates.

Agency

Mathematicians can rest easy for the moment. There is no serious indication that AI is ready to assume the finer aspects of human thinking. For mathematical intelligence to be fully realised, our approach to AI must be premised on conscious intent that is still lacking in mainstream AI applications.

Narratives around AI have begun to make space for a decoupling between intelligence and consciousness.[3] If a machine can supersede the smartest humans in the most complex tasks, then what does it matter whether they are capable of having subjective experiences?

Mathematics, the entire enterprise of which is driven by our conscious choices, tells us that it matters a great deal. If there is one thing to be learned from the histories of mathematics, it is that intellectual inquiry has no fixed trajectory. The greatest gift of all those bestowed by mathematical intelligence is agency – the freedom to think for ourselves and to actively guide ourselves through the different stages of discovery.

There is no single body of knowledge waiting to be unearthed one mechanised leap at a time – the *incompleteness* of mathematics renders it so wonderfully subjective that no single exploration can lead us to all truths. Our mathematical inquiry is tied to our decisions – the axioms we build from, the rules we follow, the questions we ponder. Each of us brings the baggage of our environment and upbringing, our language and education, in shaping our personal visions of mathematics. The scope for exploration is as vast as humanity is diverse. Mathematical thinkers set their own rules for engagement, and they can entertain the cognitive conflicts that lead them to break those very same rules while they dream up whole new worlds.

Without consciousness, there would be no aesthetic contemplation of our intellectual breakthroughs. DeepMind's Go-playing machines would be acknowledged for their outcomes, but there would be no moment to marvel at the elegance of their play. Where does our intellectual purpose derive from, if not a profound sense of appreciation?

Epilogue

The dominant paradigm of machine learning bypasses the need for consciousness. The reason it is so pervasive is that it has proved versatile enough to bring efficiencies and labour savings across a range of jobs and sectors. It works enough of the time, to enough of an extent, to generate hype and investment. The external pull is towards aggregate performance rather than towards conscious reflection of how to achieve fair and equitable outcomes for all.

AI raises all manner of moral, legal and ethical challenges. It dovetails with the economic logic of mass surveillance that has come to shape Big Tech.[4] Data is the coin of the realm, minted by the pattern-matching tools of machine learning. A disproportionate amount of energy is invested in predictive models that capture users' attention, anticipate their online movements and influence their purchasing habits and voting preferences. Governments are exploiting the same tools to regulate the behaviours, even the thoughts, of entire populations. In the military, autonomous weapons (the sanitised term for 'killer robots') are making their way to the battlefield, adding a new bitter flavour to proxy wars. As long as commercial and political incentives are the guiding hand of AI development, we risk letting machines (and their human creators) off the hook for the choices they make on behalf of society.

Someone needs to ask the hard questions and demand a higher level of scrutiny for automated judgements. If not the machines, then that leaves only us. Mathematical intelligence keeps our curiosity burning; we'll need it to prise open black-box systems, examine the rhyme and reason behind their choices and expose the gaps in their thinking. It is a welcome paradox of mathematical intelligence that it can alert us to the limits of mathematising the world; some concepts are just too unwieldy to specify and resolve in precise terms.

The pandemic is a painful reminder that science is no free lunch; the same theory can drive different practical ends. Where some nations have prospered by embracing scientific evidence from the outset, others have floundered as mathematical models have been wilfully misappropriated to serve ideological notions of liberty.

The story of human–machine collaboration is far from over, and its ending, too, will depend on the choices we make and the questions we ask. We will get to live out just one of AI's countless possible futures – given the frightening prospects of our digital creations, we may not be permitted a do-over if we misstep on our first attempt. Mathematical intelligence is a guidepost for getting us the cognitive allies we want: machines that work alongside us, for the purpose of human flourishing.

REFERENCES

Introduction

1. Quoted in K. Kelly, *Out of Control* (Basic Books, 1994), p. 34.

2. J. McCarthy et al., 'A proposal for the Dartmouth Summer Research Project in Artificial Intelligence' (31 August 1955). jmc.stanford.edu/articles/dartmouth/dartmouth.pdf

3. A widely cited study that analysed 702 occupations and placed 47 per cent of US employment at risk is C. B. Frey and M.A. Osborne, 'The future of employment: how susceptible are jobs to computerisation?', *Oxford Martin Programme on Technology and Employment* (17 September 2013). www.oxfordmartin.ox.ac.uk/downloads/academic/future-of-employment.pdf

4. B. Russell, 'The study of mathematics', in *Mysticism and Logic: And Other Essays* (Longman, 1919), p. 60.

5. G. H. Hardy, *A Mathematician's Apology* (Cambridge University Press, 1992), p. 84.

6. B. Sperber and H. Mercier, *The Enigma of Reason* (Penguin, 2017), p. 360.

7. M. R. Sundström, 'Seduced by numbers', *New Scientist* (28 January 2015). www.newscientist.com/article/mg22530062-800-the-maths-drive-is-like-the-sex-drive/

8. In *War and Peace*, Tolstoy invokes mathematics to present history not as a set of discrete events but as a continual process with infinitely small chains of cause and effect. The

same argument is made in D. Tammett, *Thinking in Numbers* (Hodder & Stoughton, 2012), p. 163.

9. E. Wigner, 'The unreasonable effectiveness of mathematics in the natural sciences', *Communications in Pure and Applied Mathematics*, vol. 13 [1] (1960).

10. J. Mubeen, 'I no longer understand my PhD dissertation (and what this means for mathematics education)', *Medium* (14 February 2016). medium.com/@fjmubeen/ ai-no-longer-understand-my-phd-dissertation-and-what-this-means-for-mathematics-education-1d40708f61c

11. M. McCourt, 'A brief history of mathematics education in England', *Emaths blog* (29 December 2017). emaths.co.uk/ index.php/blog/item/a-brief-history-of-mathematics-education-in-england

12. This may also go some way to explaining the 'unreasonable effectiveness' of mathematics; perhaps the subject lands itself in practical applications so often because we can't help but conceive of mathematical objects in terms of what we encounter in the real world. See D. Falk, 'What is math?', *Smithsonian Magazine* (23 September 2020).

13. Maths educator Frederick Peck notes: 'people use a variety of situationally-relevant methods for computation – strategies that recruit features of the situation into the computation, rather than strategies that abstract out those features ... Mathematics adheres in the relationship between people and setting.' F. A. Peck, 'Rejecting Platonism: recovering humanity in mathematics education', *Education Sciences*, vol. 8 [43] (2018).

14. Conrad Wolfram, a technologist and strong proponent for embedding computers in the maths curriculum, estimates that around 2,401 average lifetimes are spent on manual calculation in schools across the world each day. C. Wolfram, *The Math(s) Fix* (Wolfram Media Inc., 2020).

15. D. Bell, 'Summit report and key messages', *Maths Anxiety Summit 2018* (13 June 2018). Retrieved 31 July 2021 from www.

learnus.co.uk/Maths%20Anxiety%20Summit%202018%20
Report%20Final%202018–08–29.pdf

For a comprehensive survey of research into maths anxiety,
see A. Dowker et al., 'Mathematics anxiety: what have we
learned in 60 years?', *Frontiers in Psychology* (25 April 2016).

16. I. M. Lyons and S. L. Beilock, 'When math hurts: math anxiety
predicts pain network activation in anticipation of doing
math', *PLOS ONE*, vol. 7 [10] (2012).

17. A. Dowker, 'Children's attitudes toward maths deteriorate as
they get older', *British Psychological Society* (13 September
2018). www.bps.org.uk/news-and-policy/children's-attitudes-
towards-maths-deteriorate-they-get-older

18. As quoted in the preface of E. Frenkel, *Love and Math* (Basic
Books, 2013).

19. There are many wonderful accounts of the history of
mathematics. This brief summary is partly informed by
D. Struik, *A Concise History of Mathematics* (Dover
Publications, 2012); and B. Clegg, *Are Numbers Real?*
(Robinson, 2017), as well as online references such as *The Story
of Mathematics* (www.storyofmathematics.com).

20. For a primer on working memory, including its origins
in research, see D. Nikolić, 'The Puzzle of Working
Memory', *Sapien Labs* (17 September 2018). sapienlabs.org/
working-memory/

21. Quote from M. Napier, *Memoirs of John Napier of
Merchiston* (Franklin Classic Trades Press, 2018), p. 381.

22. L. Daston, 'Calculation and the division of labor, 1750–1950',
31st Annual Lecture of the German Historical Institute
(9 November 2017). www.ghi-dc.org/fileadmin/user_upload/
GHI_Washington/Publications/Bulletin62/9_Daston.pdf

23. As chronicled in M. L. Shetterly, *Hidden Figures* (William
Collins, 2016).

24. 'The story of the race to develop the pocket electronic

calculator', *Vintage Calculators Web Museum*. www.
vintagecalculators.com/html/the_pocket_calculator_race.html

25. A. Whitehead, *An Introduction to Mathematics* (Dover Books on Mathematics, 2017 reprint), p. 34.

26. K. Devlin, 'Calculation was the price we used to have to pay to do mathematics' (2 May 2018). devlinsangle.blogspot. com/2018/05/calculation-was-price-we-used-to-have.html

27. B. A. Toole, *Ada, The Enchantress of Numbers* (Strawberry Press, 1998), pp. 240–61.

28. A. Turing, 'Computing machinery and intelligence', *Mind*, vol. LIX [236] (October 1950), pp. 433–60.

29. J. Vincent, 'OpenAI's latest breakthrough is astonishingly powerful but still fighting its flaws', *Verge* (30 July 2020). www. theverge.com/21346343/gpt-3-explainer-openai-examples-errors-agi-potential

30. GPT-3., 'A robot wrote this entire article. Are you scared yet, human?', *Guardian* (8 September 2020). www.theguardian. com/commentisfree/2020/sep/08/robot-wrote-this-article-gpt-3

31. C. Chace, 'The impact of AI on journalism', *Forbes* (24 August 2020). www.forbes.com/sites/calumchace/2020/08/24/the-impact-of-ai-on-journalism/?sh=42fa62b22c46. For an analysis of how journalists view the prospect of automation, see F. Mayhew, 'Most journalists see AI robots as a threat to their industry: this is why they are wrong', *Press Gazette* (26 June 2020). www.pressgazette.co.uk/ai-journalism/

32. D. Muoio, 'AI experts thought a computer couldn't beat a human at Go until the year 2100', *Business Insider* (21 May 2016). www.businessinsider.com/ai-experts-were-way-off-on-when-a-computer-could-win-go-2016–3?r=US&IR=T

33. S. Strogatz, 'One giant step for a chess-playing machine', *New York Times* (26 December 2018). www.nytimes. com/2018/12/26/science/chess-artificial-intelligence.html

34. J. Schrittwieser et al., 'MuZero: mastering Go, chess, Shogi and Atari without rules', *DeepMind blog*

(23 December 2020). deepmind.com/blog/article/
muzero-mastering-go-chess-shogi-and-atari-without-rules

35. G. Lample and F. Charton, 'Deep learning for symbolic
    mathematics', *arXiv: 1912.01412* (2 December 2019).

36. K. Cobbe et al., 'Training verifiers to solve math word
    problems', *arXiv: 2110.14168v1* (27 October 2021).

37. A. Kharpal, 'Stephen Hawking says AI could be "worst event
    in the history of our civilisation"', *CNBC* (6 November 2017).
    www.cnbc.com/2017/11/06/stephen-hawking-ai-could-be-worst-
    event-in-civilization.html

38. The term 'superintelligence' has become mainstream thanks
    largely to N. Bostrum, *Superintelligence* (Oxford University
    Press, 2014). It is defined by Bostrom as 'an intellect that is
    much smarter than the best human brains in practically every
    field, including scientific creativity, general wisdom and social
    skills' (p. 22).

39. R. Thornton, 'The age of machinery', *Primitive Expounder*
    (1847), p. 281.

40. G. Marcus, 'Deep learning: a critical appraisal', *arXiv:
    1801.00631* (2 January 2018), provides ten critiques of deep
    learning, ranging from the data-hungry nature of these
    algorithms to their inability to distinguish correlation from
    causation.

41. M. Hutson, 'AI researchers allege that machine learning
    is alchemy', *Science* (3 May 2018). www.sciencemag.org/
    news/2018/05/ai-researchers-allege-machine-learning-alchemy

42. F. Chollet, 'The limitations of deep learning', *The Keras Blog*
    (17 July 2017). blog.keras.io/the-limitations-of-deep-learning.
    html

43. J. von Neumann, *The Computer and the Brain* (Yale University
    Press, 1958).

44. G. Zarkadakis, *In Our Own Image* (Pegasus Books, 2017),
    chronicles the metaphors of the human brain that have
    been employed through history: the biblical depiction of

man as clay/dirt imbued with God's spirit, hydraulics, automata powered by springs and gears, complex machinery, electromechanical devices (e.g. the telegraph) and computers.

45. Zoologist Matthew Cobb notes that the computational metaphor of the brain is reinforced by the fact that it can be stimulated with electrical inputs. Cobb also notes that the metaphor has been deployed in reverse: in the nineteenth century, Morse code was often described in terms of the human nervous system, and the von Neumann architecture that underpins today's computers was shaped by John von Neumann's understanding of the biological structure and function of the human brain. See the introduction to M. Cobb, *The Idea of the Brain* (Profile Books, 2020).

46. E. A. Maguire et al., 'Navigation-related structural change in the hippocampi of taxi drivers', *Proceedings of the National Academy of Sciences*, vol. 97 [8] (2000), pp. 4398–403.

47. This idea is explored in detail in D. Eagleman, *Livewired* (Canongate Books, 2020).

48. A. Newell et al., 'Chess-playing programs and the problem of complexity', *IBM Journal of Research and Development* (4 October 1958). ieeexplore.ieee.org/document/5392645

49. B. Weber, 'Mean chess-playing computer tears at meaning of thought', *New York Times* (19 February 1996). besser.tsoa.nyu. edu/impact/w96/News/News7/0219weber.html

50. The AlphaFold Team, 'AlphaFold: a solution to a 50-year-old grand challenge in biology', *DeepMind Blog* (30 November 2020). deepmind.com/blog/article/alphafold-a-solution-to-a-50-year-old-grand-challenge-in-biology

    In July 2021, DeepMind made AlphaGo open access, and also published a database of hundreds of thousands of 3D protein structures, available (with more to follow) at alphafold. ebi.ac.uk

51. G. Kasparov, 'The chess master and the computer', *New York*

Review of Books (11 February 2010). www.nybooks.com/
articles/2010/02/11/the-chess-master-and-the-computer/

52. H. Moravec, *Mind Children* (Harvard University Press, 1988),
    p. 15.

53. R. Sennett, *The Craftsman* (Penguin, 2009), p. 105.

54. V. Kramnik, 'Vladimir Kramnik on man vs machine',
    ChessBase (18 December 2018). en.chessbase.com/post/
    vladimir-kramnik-on-man-vs-machine

55. G. Kasparov, *Deep Thinking* (John Murray, 2017), p. 246.

56. D. Susskind, *A World Without Work* (Allen Lane, 2020),
    defines the Age of Labour as the historical period in which the
    substituting force of technology is offset by its complementary
    force, resulting in new forms of employment. Susskind argues
    that the Age of Labour is at an end because the substituting
    force of machine learning technologies are overwhelming their
    complementary force, to the detriment of future job prospects.

57. An argument Kasparov himself has made – see G. Kasparov,
    'Chess, a *Drosophila* of reasoning', *Science* (7 December 2018).
    science.sciencemag.org/content/362/6419/1087.full

58. An outline of the approach to the proof, and the many iterative
    attempts leading up to it, is presented in R. Wilson, *Four
    Colors Suffice* (Princeton University Press, 2013).

59. A. Davies et al., 'Advancing mathematics by guiding human
    intuition with AI', *Nature*, 600 (2021), pp. 70–74. doi.
    org/10.1038/s41586-021-04086-x
    One of the mathematicians involved in the paper
    shares his reflections on the collaborative potential of AI:
    G. Williamson, 'Mathematical discoveries take intuition
    and creativity – and now a little help from AI', *The
    Conversation* (1 December 2021). theconversation.com/
    mathematical-discoveries-take-intuition-and-creativity-and-
    now-a-little-help-from-ai-172900?utm_source=pocket_mylist

60. A. Clark and D. J. Chalmers, 'The extended mind', *Analysis*,
    vol. 58 [1] (1998), pp. 7–19.

61. Geoff Hinton, a pioneer of deep learning, admits readily of these algorithms that 'we really don't know how they work'. N. Thompson, 'An AI pioneer explains the evolution of neural networks', *Wired* (13 May 2019). www.wired.com/story/ai-pioneer-explains-evolution-neural-networks/

62. C. O'Neil, *Weapons of Math Destruction* (Penguin, 2016).

63. See, for example, K. Crawford, 'Artificial intelligence's white guy problem', *New York Times* (25 June 2016). www.nytimes.com/2016/06/26/opinion/sunday/artificial-intelligences-white-guy-problem.html

64. J. Dastin, 'Amazon scraps secret AI recruiting tool that showed bias against women', *Reuters* (10 October 2018). www.reuters.com/article/us-amazon-com-jobs-automation-insight-idUSKCN1MK08G

65. T. Simonite, 'Photo algorithms ID white men fine – black women, not so much', *Wired* (2 June 2018). www.wired.com/story/photo-algorithms-id-white-men-fineblack-women-not-so-much/ and T. Simonite, 'When it comes to gorillas, Google Photos remains blind', *Wired* (1 November 2018). www.wired.com/story/when-it-comes-to-gorillas-google-photos-remains-blind/

66. L. Howell, 'Digital wildfires in a hyperconnected world', *World Economic Forum Report*, vol. 3 (2013), pp. 15–94.

Chapter 1. Estimation

1. L. Ostlere, 'VAR arrives in the Premier League to unearth sins we didn't know existed', *Independent* (14 August 2019). www.independent.co.uk/sport/football/premier-league/var-premier-league-offside-raheem-sterling-bundesliga-mls-clear-and-obvious-a9056906.html

   The accuracy of VAR is not above reproach; its reliance on camera frames means that decisions carry a margin of error, as football writer Jonathan Wilson explained in a Twitter thread: twitter.com/jonawils/status/1160241782506086401

2. Everett's accounts of the Pirahâ are documented in D. Everett, *Don't Sleep, There Are Snakes* (Profile Books, 2010). See also J. Colapinto, 'The Interpreter', *New Yorker* (9 April 2007). www.newyorker.com/magazine/2007/04/16/the-interpreter-2

3. N. Chomsky, 'Things no amount of learning can teach', *Noam Chomsky interviewed by John Gliedman* (November 1983). chomsky.info/198311__/

4. J. Gay, 'Mathematics among the Kpelle Tribe of Liberia: preliminary report', *African Education Program (Educational Services Incorporated)* (1964). Retrieved 31 July 2021 from lchcautobio.ucsd.edu/wp-content/uploads/2015/10/Gay-1964-Math-among-the-Kpelle-Ch-1.pdf

5. The findings in this section are described at length in K. Devlin, *The Math Instinct* (Thunder's Mouth Express, 2005), Chapter 1; and S. Dehaene, *The Number Sense* (Oxford University Press, 2011). For the original Karen Wynn experiment, see K. Wynn, 'Addition and subtraction by human infants', *Nature*, vol. 358 (1992), pp. 749–50.

6. S. Dehaene, *The Number Sense* (Oxford University Press, 2011).

7. J. Holt, 'Numbers guy', *New Yorker* (25 February 2008). www.newyorker.com/magazine/2008/03/03/numbers-guy

8. L. Feigenson et al., 'Core systems of number', *Trends in Cognitive Sciences*, vol. 8 [7] (2004), pp. 307–14.

9. These examples fall under the philosophical category of 'vagueness', as defined by *Stanford Encyclopedia of Philosophy Archive* (5 April 2018). plato.stanford.edu/archives/sum2018/entries/vagueness/

10. A. Starr et al., 'Number sense in infancy predicts mathematical abilities in childhood', *Proceedings of the National Academy of Sciences*, vol. 110 [45] (2013), pp. 18116–20. www.pnas.org/content/110/45/18116

11. R. S. Siegler and J. L. Booth, 'Development of numerical

estimation: a review', in J. I. D. Campbell (ed.), *Handbook of Mathematical Cognition* (Psychology Press, 2005), pp. 197–212.

12. J. Boaler, *What's Math Got to Do With It?* (Viking Books, 2008), p. 25.

13. W. T. Kelvin, 'The six gateways of knowledge', *Presidential Address to the Birmingham and Midland Institute*, Birmingham, 1883 – later published in *Popular Lectures and Addresses*, vol. 1, 280, 1891.

14. E. Fermi, 'Trinity Test, July 16, 1945, eyewitness accounts', *US National Archives*, 16 July 1945. www.dannen.com/decision/fermi.html

15. 'Fermi's piano tuner problem', *NASA*, www.grc.nasa.gov/www/k-12/Numbers/Math/Mathematical_Thinking/fermis_piano_tuner.htm

16. J. Cepelewicz, 'The hard lessons of modeling the coronavirus pandemic', *Quanta Magazine* (28 January 2021). www.quantamagazine.org/the-hard-lessons-of-modeling-the-coronavirus-pandemic-20210128/

17. Example taken from K. Yates, *The Maths of Life and Death* (Quercus, 2019), Chapter 4. The system used the last two digits of the current year and year of birth to calculate each patient's age. So in the year 2000, patients born in 1965 were assigned an age of −35.

18. O. Solon, 'How a book about flies came to be priced $24 million on Amazon', *Wired* (27 April 2011). www.wired.com/2011/04/amazon-flies-24-million/

19. J. Earl, '6-year-old orders $160 dollhouse, 4 pounds of cookies with Amazon's Echo Dot', *CBS News* (5 January 2017). www.cbsnews.com/news/6-year-old-brooke-neitzel-orders-dollhouse-cookies-with-amazon-echo-dot-alexa/

20. K. Campbell-Dollaghan, 'This neural network is hilariously bad at describing outer space', *Gizmodo* (19 August 2015). gizmodo.com/this-neural-network-is-hilariously-bad-at-describing-ou-1725195868

References

21. J. Vincent, 'Twitter taught Microsoft's AI chatbot to be a racist asshole in less than a day', *Verge* (24 March 2016). www.theverge.com/2016/3/24/11297050/tay-microsoft-chatbot-racist

22. B. Finio, 'Measure Earth's circumference with a shadow', *Scientific American* (7 September 2017). www.scientificamerican.com/article/measure-earths-circumference-with-a-shadow/

23. V. F. Rickey, 'How Columbus encountered America', *Mathematics Magazine*, vol. 65 [4] (1992), pp. 219–225.

24. C. G. Northcutt et al., 'Pervasive label errors in test sets destabilize machine learning benchmarks', *arXiv.org* (26 March 2021). arxiv.org/abs/2103.14749

25. G. Press, 'Andrew Ng launches a campaign for data-centric AI', *Forbes* (16 June 2021). www.forbes.com/sites/gilpress/2021/06/16/andrew-ng-launches-a-campaign-for-data-centric-ai/?sh=3b02f1674f57

26. For example: xkcd.com/2205/

27. F. I. M. Craik and J. F. Hay, 'Aging and judgments of duration: effects of task complexity and method of estimation', *Perceptions & Psychophysics*, vol. 61 [3] (1999), pp. 549–60. link.springer.com/article/10.3758/BF03211972

28. Example taken from 'Why logarithms still make sense', *Chalkface Blog* (7 March 2016). thechalkfaceblog.wordpress.com/2016/03/07/why-logarithms-still-make-sense/

29. D. Robson, 'Exponential growth bias: the numerical error behind Covid-19', *BBC Future* (14 August 2020). www.bbc.com/future/article/20200812-exponential-growth-bias-the-numerical-error-behind-covid-19

30. G. S. Goda et al., 'The role of time preferences and exponential-growth bias in retirement savings', *National Bureau of Economic Research, Working Paper 21482*, August 2015.

31. R. Banerjee et al., 'Exponential-growth prediction bias and

compliance with safety measures in the times of Covid-19', *IZA Institute of Labor Economics* (May 2020).

32. A. Romano et al., 'The public do not understand logarithmic graphs used to portray Covid-19', *LSE blog* (19 May 2020). blogs.lse.ac.uk/covid19/2020/05/19/the-public-doesnt-understand-logarithmic-graphs-often-used-to-portray-covid-19/

    For a comparison of linear and logarithmic scales concerning UK Covid-19 cases, see 'Exponential growth: what it is, why it matters, and how to stop it', *Centre for Evidence-Based Medicine (University of Oxford)* (23 September 2020). www.cebm.net/covid-19/exponential-growth-what-it-is-why-it-matters-and-how-to-spot-it/

33. As quoted in S. Radcliffe, 'Roy Amara 1925–2007, American futurologist', in *Oxford Essential Quotations (4th ed.)* (Oxford University Press, 2016).

34. M. Schonger and D. Sele, 'How to better communicate exponential growth of infectious diseases', *PLOS ONE,* vol. 15 [12] (2020).

35. J. Searle, 'Minds, brains, and programs', *Behavioral and Brain Sciences*, vol. 3 [3] (1980), pp. 417–57.

36. K. Reusser, 'Problem solving beyond the logic of things: contextual effects on understanding and solving word problems', *Instructional Science*, vol. 17 [4] (1988), pp. 309–338.

37. See L. Chittka L. and K. Geiger, 'Can honey bees count landmarks?', *Animal Behaviour*, vol. 49 [1] (1995), pp. 159–64; K. McComb et al., 'Roaring and numerical assessment in contests between groups of female lions, *Panthera leo*', *Animal Behaviour*, vol. 47 [2] (1994), pp. 379–87; R. L. Rodríguez et al., '*Nephila clavipes* spiders (Araneae: Nephilidae) keep track of captured prey counts: testing for a sense of numerosity in an orb-weaver', *Animal Cognition*, vol. 18 (2015), pp. 307–14; M. E. Kirschhock et al., 'Behavioral and neuronal representation of numerosity zero in the crow', *Journal of Neuroscience*, vol. 41 [22] (2021), pp. 4889–96.

38. Rats and chimpanzees are among the animals covered in K. Devlin, *The Math Instinct* (Thunder's Mouth Express, 2005). Extensions of Karen Wynn's findings to rhesus monkeys are cited in S. Dehaene, *The Number Sense* (Oxford University Press, 2011), pp. 53–5. For extensions to dogs, see R. West and R. Young, 'Do domestic dogs show any evidence of being able to count?', *Animal Cognition*, vol. 5 [3] (2002), pp. 183–6.

Chapter 2. Representation

1. The General Problem Solver is introduced in A. Newell et al., 'Report on a general problem-solving program', *Proceedings of the International Conference on Information Processing* (1959), pp. 256–64. The GPS relied on means-end analysis: a goal state is specified, and actions are taken to reduce the gap between the current state and the goal state. For example, if the goal state is to restock your empty fridge with milk, then the GPS might choose to take a trip to the supermarket, which gets you closer towards the final goal. The GPS relied on searching through possible actions; for more open problems, the sheer size of the search space proved prohibitive.

2. For an early account of the limitations of rule-based systems see H. Dreyfus, *What Machines Can't Do* (MIT Press, 1972). Dreyfus refuted four assumptions that underpinned early approaches to AI: 1) the biological assumption (the brain processes information via some biological equivalent of on/off switches), 2) the psychological assumption (the mind is a device that operates on bits of information according to certain formal rules), 3) the epistemological assumption (all knowledge can be formalised) and 4) the ontological assumption (the world can be described by a collection of facts, each represented by a symbol).

3. M. Polanyi, *The Tacit Dimension* (University of Chicago Press, 1966), p. 4.

4. A neural network consists of several layers of 'neurons', with

the inputs fed into the bottom layer, and the output at the top layer. Each neuron connects with thousands of its neighbours, with the strength of the connections determined by weights. The 'deep' part of this so-called 'deep learning' simply corresponds to there being many layers, which requires a degree of computing power that has only become available in the last couple of decades.

5. In the case of AlphaGo, a neural network was deployed to reduce the search space to a more manageable size by flagging up the most effective moves. For each of those possible moves, AlphaGo then simulated what might happen in the rest of the game several times over and used the results to pick out the best move among them. There is a lot of search happening at this stage, and another neural network acts as a guide to eliminate all but the most promising choices.

AlphaGo began by examining a database of around 30 million human Go moves and, after attaining some proficiency in the game, it then played other instances of itself over and over. In every simulated game, AlphaGo learned to value every move that improved its game position and to devalue those that did not (using *reinforcement learning*). It then used *deep learning* to assess configurations on the board and decide which features of gameplay correlated to better performance.

Rather than explicitly being told how to play well, as Deep Blue was, AlphaGo was given the ground rules of the game and studied games played by humans to figure out for itself the most effective moves. DeepMind has since gone further by developing successors to AlphaGo that learn purely from data they generate by playing themselves. No human gameplay is required to bootstrap their capabilities.

See D. Silver et al., 'Mastering the game of Go with deep neural networks and tree search', *Nature*, vol. 529 (2016), pp. 484–9. www.nature.com/articles/nature16961

AlphaGo's immediate successor, AlphaGo Zero, which

took no input from human games (it learns strategies through self-play alone), earned an emphatic 100–0 victory over AlphaGo, intensifying claims of its superhuman skill: 'AlphaGo Zero: starting from scratch', *DeepMind blog* (18 October 2017). deepmind.com/blog/article/alphago-zero-starting-scratch

6. This is the crux of Jeff Hawkins' Thousand Brains Theory, which posits that the neocortex learns models of the world using maplike reference frames, which are stored in thousands of cortical columns – see J. Hawkins, *A Thousand Brains: A New Theory of Intelligence* (Basic Books, 2021).

7. H. Fry, *Hello World* (Transworld Digital, 2018), Chapter 4 (Medicine).

8. M. T. Ribiero et al., '"Why should I trust you?": explaining the predictions of any classifier', in *Proceedings of the 22nd ACM SIGKDD International Conference on Knowledge Discovery and Data Mining* (August 2016), pp. 1135–44.

9. J. K. Winkler et al., 'Association between surgical skin markings in dermoscopic images and diagnostic performance of a deep learning convolutional neural network for melanoma recognition', *JAMA Dermatology*, vol. 155 [10] (2019), pp. 1135–41.

10. W. Samek et al., 'Explaining deep neural networks and beyond: a review of methods and applications', *Proceedings of the IEEE*, vol. 109 [3] (2021), pp. 247–78.

11. T. B. Brown et al., 'Adversarial patch', *arXiv.org* (17 May 2018). arxiv.org/pdf/1712.09665.pdf

12. G. Marcus and D. Ernest, 'GPT-3, Bloviator: OpenAI's language generator has no idea what it's talking about', *MIT Technology Review* (22 August 2020). www.technologyreview.com/2020/08/22/1007539/gpt3-openai-language-generator-artificial-intelligence-ai-opinion/

13. For example, G. Marcus, 'The next decade in AI: four steps towards robust artificial intelligence', *arXiv.org* (17 February 2020).

14. S. Dehaene, *How We Learn* (Allen Lane, 2020), p. 15. See also A. M. Zador, 'A critique of pure learning and what artificial neural networks can learn from animal brains', *Nature Communications*, vol. 10 (August 2019), which argues that our brain's wiring is far too complex for the genome to specify explicitly. Zador introduces the idea of a *genomic bottleneck* – 'the compression into the genome of whatever innate processes are captured by evolution'.

15. S. Dehaene, *How We Learn* (Allen Lane, 2020), p. 17.

16. 'Is there a better way to count …? 12s anyone?', *Angel Sharp Media* (28 September 2018). www.bbc.com/ideas/videos/ is-there-a-better-way-to-count-12s-anyone/p06mdfkn

17. Stephen Wolfram argues, for instance, that numbers are not an inevitable construct for intelligent beings. He suggests they are an artefact of how humans observe the universe and perform computations. S. Wolfram, 'How inevitable is the concept of numbers?', *Stephen Wolfram Writings Blog* (25 May 2021). writings.stephenwolfram.com/2021/05/ how-inevitable-is-the-concept-of-numbers/

18. T. Landauer, 'How much do people remember? Some estimates of the quantity of learned information in long-term memory', *Cognitive Science*, vol. 10 [4] (1986), pp. 477–93. www. cs.colorado.edu/~mozer/Teaching/syllabi/7782/readings/ Landauer1986.pdf

19. This figure is derived from an anatomical estimate of the capacity of each synapse, which is placed at around 4.7 bits of information. See T. M. Bartol et al., 'Nanoconnectomic upper bound on the variability of synaptic plasticity', *eLife*, vol. 4 [e10778] (2015).

20. The Hutter Prize, for instance, offers €5,000 to anyone who can design a program that compresses a 1GB English text file with more than 1 per cent improvement on the latest record. The organisers are hoping to encourage progress towards Artificial

General Intelligence (AGI), which they believe text compression is equivalent to. See prize.hutter1.net

21. The cognitive scientist Donald Hoffman has gone further by arguing that our perceptions of the world are just an interface that has arisen through natural selection to help inform us of fitness payoffs. The true nature of reality, in this view, is concealed behind the arbitrary formats we have adopted to process the world. For this argument, and references to the figures quoted in this paragraph, see D. Hoffman, *The Case Against Reality* (Penguin, 2019).

22. A. Ericsson, *Peak* (Vintage, 2017), pp. 60–61.

23. A. D. de Groot, 'Het denken van de schaker', *PhD dissertation* (1946). Translated into *Thought and Choice in Chess* (Mouton Publishers, 1965).

24. W. G. Chase and H. A. Simon, 'Perception in chess', *Cognitive Psychology*, vol. 4 [1] (1973), pp. 55–81.

25. E. Cooke, 'Let a grandmaster of memory teach you something you'll never forget', *Guardian* (7 November 2015). www.theguardian.com/education/2015/nov/07/grandmaster-memory-teach-something-never-forget

26. As cited in M. F. Dahlstrom, 'Using narratives and storytelling to communicate science with nonexpert audiences', *Proceedings of the National Academy of Sciences*, vol. 111 [4] (2014), pp. 13614–20.

27. W. P. Thurston, 'Mathematical education', *Notices of the American Mathematical Society* (1990), pp. 844–50.

28. For a mathematical argument in favour of restricting multiplication tables to 10 × 10, on the basis of a trade-off between memorisation effort and approximation benefits for large calculations, see J. Mcloone, 'Is there any point to the 12 times table?', *Wolfram Blog* (26 June 2013). blog.wolfram.com/2013/06/26/is-there-any-point-to-the-12-times-table/

29. H. Poincaré, 'Hypotheses in physics' (Chapter 9) in *Science and Hypothesis* (Walter Scott Publishing, 1905), pp. 140–59.

30. From 'The true scale multiplication grid', *Chalkface Blog* (29 April 2017). thechalkfaceblog.wordpress.com/2017/04/29/the-true-scale-multiplication-grid/

31. Image from M. Watkins, *Secrets of Creation, Volume 1: The Mystery of Prime Numbers* (Liberalis, 2015), p. 66, the inspiration of which is an approach laid out in A. Doxiadis, *Uncle Petros and Goldbach's Conjecture* (Faber & Faber, 2001).

32. For a thorough treatment of arithmetic predicated on symbol knitting see P. Lockhart, *Arithmetic* (Harvard University Press, 2019).

33. S. Zeki et al., 'The experience of mathematical beauty and its neural correlates', *Frontiers in Human Neuroscience* (13 February 2014). www.frontiersin.org/articles/10.3389/fnhum.2014.00068/full

34. Grant Sanderson, creator of popular maths video channel *3Blue1Brown*, notes that the notation $e^{i\pi}$ is somewhat misleading because it suggests that this term has to do with powers and repeated multiplication when, in fact, it is based on a particular definition of the exponential function as applied to complex numbers. On that basis, he suggests the beauty of Euler's formula has been overstated. See G. Sanderson, 'What is Euler's formula actually saying?', *3Blue1Brown YouTube Channel* (28 April 2020). www.youtube.com/watch?v=ZxYOEwM6Wbk.

    Others have argued that the beauty is retained and appreciated in a new light once the deeper mathematical interpretation of the equation is understood. See L. Devlin, 'Is Euler's Identity beautiful? And if so, how?', *Devlin's Angle (Mathematical Association of America)* (June 2021). www.mathvalues.org/masterblog/is-eulers-identity-beautiful-and-if-so-how

35. For a history of mathematical notation see J. Mazur, *Enlightening Symbols* (Princeton University Press, 2014).

36. K. Devlin, 'Algebraic roots – Part 1', *Devlin's Angle* (4 April

2016). devlinsangle.blogspot.com/2016/04/algebraic-roots-part-1.html

37. Despite some attempts (for example, S. B. Sells and R. S. Fixott, 'Evaluation of research on effects of visual training on visual functions', *American Journal of Ophthalmology,* vol. 44 [2] (1957), pp. 230–36), it is difficult to quantify how much of the brain is dedicated to vision, since there is a lot of overlap between the senses, and brain regions tend to be multimodal.

38. V. Menon, 'Arithmetic in child and adult brain', in K. R. Cohen and A. Dowker, *Handbook of Mathematical Cognition* (Oxford University Press, 2014). doi.org/doi:10.1093/oxfordhb/9780199642342.013.041

39. J. Boaler et al., 'Seeing as understanding: the importance of visual mathematics for our brain and learning', *youcubed* (March 2017). www.youcubed.org/wp-content/uploads/2017/03/Visual-Math-Paper-vF.pdf

40. Music Animation Machine: musanim.com

41. H. Jacobson, 'The world has lost a great artist in mathematician Maryam Mirzakhani', *Guardian* (29 July 2017). www.theguardian.com/science/2017/jul/29/maryam-mirzakhani-great-artist-mathematician-fields-medal-howard-jacobson

42. E. Klarreich, 'Meet the first woman to win math's most prestigious prize', *Wired* (13 August 2014). www.wired.com/2014/08/maryam-mirzakhani-fields-medal/

43. S. Russell, *Human Compatible* (Allen Lane, 2019), p. 81.

44. The radiation problem dates back to Karl Duncker's 1945 study: K. Duncker, 'On problem solving', *Psychological Monographs,* vol. 58 [270] (1945). Conditions under which the success rate increases following exposure to analogy are covered in M. L. Gick and K. J. Holyoak, 'Analogical problem solving', *Cognitive Psychology,* vol. 12 (1980), pp. 306–55, and M. L. Gick and K. J. Holyoak, 'Schema introduction and analogical transfer, *Cognitive Psychology,* vol. 15 (1983), pp. 1–38.

45. J. Pavlus, 'The computer scientist training AI to think with

analogies', *Quanta Magazine* (14 July 2021). www.
quantamagazine.org/melanie-mitchell-trains-ai-to-think-with-
analogies-20210714/

46. M. Atiyah, 'Identifying progress in mathematics', *The
Identification of Progress in Learning* (Cambridge University
Press, 1985), pp. 24–41.

47. A. Sierpinska, 'Some remarks on understanding in
mathematics', *Canadian Mathematics Education Study Group*
(1990). flm-journal.org/Articles/43489F40454C8B2E06F334CC1
3CCA8.pdf

48. For an elaboration of these differences see A. Cuoco et al.,
'Habits of mind: an organizing principle of mathematics
curricula', *Journal of Mathematical Behaviour* (December
1996), pp. 375–402.

49. As quoted in E. Frenkel, *Love and Math* (Basic Books, 2013),
Preface.

50. T. N. Carraher et al., 'Mathematics in the street and in school',
British Journal of Developmental Psychology, vol. 3, [1]
(1985), pp. 21–29 and G. Saxe, 'The mathematics of child street
vendors', *Child Development*, vol. 59 [5] (1988), pp. 1415–25.

51. H. Fry, 'What data can't do', *New Yorker* (29 March 2021).
www.newyorker.com/magazine/2021/03/29/what-data-cant-do

52. P. Vamplew, 'Lego Mindstorms robots as a platform for
teaching reinforcement learning', *Proceedings of AISAT2004:
International Conference on Artificial Intelligence in Science
and Technology* (2004) and P. Vamplew et al., 'Human-aligned
artificial intelligence is a multiobjective problem', *Ethics and
Information Technology*, vol. 20, [1] (2018), pp. 27–40.

53. Norbert Wiener was among the first to speculate on possible
misalignment between humans and machines, noting in 1960
that 'we had better be quite sure that the purpose put into the
machine is the purpose which we really desire.' N. Wiener,
'Some moral and technical consequences of automation',
Science, vol. 131 [3410] (1960), pp. 1355–8. For a contemporary

treatment of the value alignment problem, and strategies for solving it, see S. Russell, *Human Compatible* (Penguin, 2019).

54. A. Pasick, 'Here are some of the terrifying possibilities that have Elon Musk worried about artificial intelligence', *Quartz* (4 August 2014). qz.com/244334/here-are-some-of-the-terrifying-possibilities-that-have-elon-musk-worried-about-artificial-intelligence/

55. The 'inner world of human life', according to mathematicians Phillip Davis and Reuben Hersh, includes things such as our emotions, attitudes and intentions, and 'can never be mathematized'. See P. J. Davis and R. Hersh, *Descartes' Dream: The World According to Mathematics* (Penguin, 1988), p. 23.

Chapter 3. Reasoning

1. Bertrand Russell first gave the example of the ill-fated Thanksgiving chicken in his 1912 book *The Problems of Philosophy*. It was adapted to the case of the Christmas turkey by Karl Popper in A. Chalmers, *What is This Thing Called Science?* (University of Queensland Press, 1982).

2. For a worked solution see 'Circle division solution' on 3Blue1Brown's YouTube channel: www.youtube.com/watch?v=K8P8uFahAgc&t=188s

3. With purely data-driven prediction models, there is always a risk of stretching conclusions too far. A mathematical result called the 'No-Free Lunch Theorem' says that given any machine learning algorithm and half of a given dataset, it is always possible to tinker with the remaining half – the unseen data – in such a way that the algorithm makes good predictions on the training data but misfires on the unseen data. The damning implication is that no single algorithm can predict outcomes accurately under any scenario – the only guarantee is in memorising the past.

4. This explains why we are so vulnerable to (and fascinated by) optical illusions – see B. Resnick, '"Reality" is constructed by

your brain. Here's what that means and why it matters', *Vox* (22 June 2020). www.vox.com/science-and-health/20978285/ optical-illusion-science-humility-reality-polarization

5. D. Eagleman, 'The moral of the story', *New York Times* (3 August 2012). www.nytimes.com/2012/08/05/books/review/ the-storytelling-animal-by-jonathan-gottschall.html

6. The term 'Interpreter' is coined in M. Gazzaniga, *Who's in Charge?* (Robinson, 2012), p. 75.

Gazzaniga ran a famous experiment on a split-brain patient, an epileptic who, as part of his treatment, had the wiring between his two hemispheres severed. This meant that Gazzaniga could expose each hemisphere to different pictures and see how the brain made sense of it all. First an image of a chicken claw was shown to the patient's right visual field, which corresponds to the brain's left hemisphere. Then a snow scene was shown to the left visual field (right hemisphere). The patient was then shown an array of pictures, placed in full view so that both hemispheres could see. The patient pointed to a chicken with his right hand, which was hardly surprising given that his right visual field had seen a chicken claw. He also pointed to a shovel with his left hand – also to be expected, given that the snow scene was shown to the left visual field.

Next the patient was asked to explain his choices, which is where things get interesting. Summoning the speech centre of his left hemisphere, the patient replied that the chicken claw goes with the chicken. Then – here comes the twist – he looked at the shovel and explained: 'You need a shovel to clean out the chicken shed.' Remember, the left hemisphere knew nothing of the snow scene. Rather than humbly confessing 'I don't know', the patient, guided by his left hemisphere, invented a plausible story to fill gaps in his memory (chickens do make mess, after all) – plausible, but fictitious.

See also M. Gazzaniga, 'The storyteller in your head',

Discover Magazine (1 March 2012). www.discovermagazine.
com/mind/the-storyteller-in-your-head

7. D. Kahneman, *Thinking, Fast and Slow* (Penguin, 2012).

8. Quotes taken from the introduction of J. Haidt, *The Righteous Mind* (Penguin, 2013).

9. D. Sperber and H. Mercier, *The Enigma of Reason* (Penguin, 2017).

10. D. Hume, *A Treatise of Human Nature* (1739). See www.pitt.
edu/~mthompso/readings/hume.influencing.pdf

11. A. Damasio, *Descartes' Error* (Vintage, 2006).

12. F. Heider and M. Simmel, 'An experimental study of apparent behavior', *American Journal of Psychology*, vol. 57 (1944), pp. 243–59. The animation can be found on YouTube: 'Heider and Simmel movie'. www.youtube.com/watch?v=76p64j3H1Ng

13. K. Zunda, 'The case for motivated reasoning', *Psychological Bulletin*, vol. 108 [3] (1990), pp. 480–98.

14. S. Martinez-Conde et al. gives the example of a well-known trick of the magician Teller based on fake coin tosses. Here, we mistakenly associate the sound of a coin clink with an action Teller wants us to believe is occurring (new coins being plucked out and dropped) but which actually isn't. As the authors put it: 'When A precedes B, we conclude that A caused B.' S. Martinez-Conde et al., *Sleights of Mind* (Profile Books, 2012), p. 192.

15. R. Epstein and R. E. Robertson, 'The search engine manipulation effect (SEME) and its possible impact on the outcomes of elections', *Proceedings of the National Academy of Sciences*, vol. 112 [33], E4512–21 (2015). www.pnas.org/
content/112/33/E4512

16. A good summary of the affair can be found at D. Kolkman, 'F*ck the algorithm? What the world can learn from the UK's A-Level grading fiasco', *LSE blog* (26 August 2020). blogs.lse.
ac.uk/impactofsocialsciences/2020/08/26/fk-the-algorithm-
what-the-world-can-learn-from-the-uks-a-level-grading-fiasco/

17. T. Harford, 'Don't rely on algorithms to make life-changing decisions', *Financial Times* (21 August 2020). www.ft.com/content/f32b3124–6b77–4b33–9de1–7dbc6599724b

18. H. Stewart, 'Boris Johnson blames "mutant algorithm" for exams fiasco', *Guardian* (26 August 2020). www.theguardian.com/politics/2020/aug/26/boris-johnson-blames-mutant-algorithm-for-exams-fiasco

19. In J. Pearl, *The Book of Why* (Penguin, 2019) p.28, Pearl describes three levels of abstraction: Level 1, the lowest, is Association ('observing', 'seeing'), Level 2 is Intervention ('doing') and Level 3 is Counterfactuals ('Imagining', 'Retrospection', 'Understanding').

20. According to a 2019 survey, 40 per cent of companies worldwide use AI in some way to screen candidates. See D. W. Brin, 'Employers embrace artificial intelligence for HR', *SHRM*, vol. 22 (March 2019).

21. J. Dastin, 'Amazon scraps secret AI recruiting tool that showed bias against women', *Reuters* (11 October 2018). www.reuters.com/article/us-amazon-com-jobs-automation-insight-idUSKCN1MK08G

22. Tyler Vigen hosts an entertaining collection of spurious correlations at www.tylervigen.com/spurious-correlations

23. K. Crawford et al., 'AI Now 2019 Report', *AI Institute* (December 2019). ainowinstitute.org/AI_Now_2019_Report.pdf

24. 'Processes of special categories of personal data', *Article 9 of EU GDPR* (25 May 2018). www.privacy-regulation.eu/en/article-9-processing-of-special-categories-of-personal-data-GDPR.htm

25. Misinformation is rampant because divisive content spreads more rapidly than neutral content, sources are not always easy to trace (and therefore escape scrutiny) and people are willing to spread such content that reinforces their beliefs ('confirmation bias'). See H. Rahman, 'Why are social

media platforms still so bad at combating misinformation?', *KelloggInsight* (3 August 2020). insight.kellogg.northwestern.edu/article/social-media-platforms-combating-misinformation

26. H. Arendt, 'Truth and politics', *New Yorker* (25 February 1967).

27. I. Sample, 'Study blames YouTube for rise in number of flat earthers', *Guardian* (17 February 2019). www.theguardian.com/science/2019/feb/17/study-blames-youtube-for-rise-in-number-of-flat-earthers

28. W. J. Brady et al., 'Emotion shapes the diffusion of moralized content in social networks', *Proceedings of the National Academy of Sciences*, vol. 114 [28] (2017), pp. 7313–18.

29. E. Newman et al., 'Truthiness and falsiness of trivial claims depend on judgmental contexts', *Journal of Experimental Psychology Learning, Memory and Cognition*, vol. 41 [5] (2015), pp. 1337–48.

30. Coined in N. Schick, *Deep Fakes and the Infocalypse* (Monoray, 2020).

31. P. Hoffman, *The Man Who Loved Only Numbers* (Fourth Estate, 1999), p. 29.

32. The idea of 'permanence', as enshrined in mathematical proof, is elaborated in G. H. Hardy, *A Mathematician's Apology* (Cambridge University Press, 1992).

33. E. Cheng, *The Art of Logic* (Profile Books, 2018), p. 12.

34. A lovely exposition of Euclid's influence on Lincoln can be found in J. Ellenberg, *Shape* (Penguin, 2021), Chapter 1.

35. For a summary of the 4,000-year history of Pythagoras' theorem (including its pre-Pythagorean origins) see B. Ratner, 'Pythagoras: everyone knows his famous theorem, but not who discovered it 1000 years before him', *Journal of Targeting, Measurement and Analysis for Marketing*, vol. 17 [3] (2009), pp. 229–42.

36. Philosopher Imre Lakatos termed this a heuristic style of learning, arguing that proofs form a tight coupling with

refutations; we arrive at definitions and theorems by first confronting our mistruths. I. Lakatos, *Proof and Refutations* (Cambridge University Press, 1976).

37. E. S. Loomis, *The Pythagorean Proposition* (Tarquin Publications, 1968).

38. See, for example, D. Mackenzie, *Mechanizing Proof* (MIT Press, 2004), which traces such efforts among software developers to the 1960s.

39. A collection of conjectures that have required really large counterexamples can be found at math.stackexchange.com/ questions/514/conjectures-that-have-been-disproved-with-extremely-large-counterexamples.

40. J. Horgan, 'The death of proof', *Scientific American*, vol. 269 [4] (1993). www.math.uh.edu/~tomforde/Articles/ DeathOfProof.pdf

41. Modern computing methods such as SAT ('satisfiability') now allow computers themselves to reduce some problems with infinite options down to a discrete, finite form. See K. Harnett, 'Computer scientists attempt to corner the Collatz conjecture', *Quanta Magazine* (26 August 2020). www.quantamagazine. org/can-computers-solve-the-collatz-conjecture-20200826/

42. In fact, Euler's conjecture was more general, stating that there are no non-zero solutions for any equation of the form
$$x_k^k = x_1^k + x_2^k + \ldots + x_{k-1}^k$$
for any k $\geq$3. As stated in W. Dunham, 'The genius of Euler: reflections on his life and work', *Mathematical Association of America*, 2007, p. 220.

43. For example, Andrew Booker used a supercomputer to prove that the number 33 can be expressed as the sum of three cubes. See J. Pavlus, 'How search algorithms are changing the course of mathematics', *Nautilus* (28 March 2019). nautil.us/issue/70/ variables/how-search-algorithms-are-changing-the-course-of-mathematics

44. K. Hartnerr, 'Building the mathematical library of the future',

Quanta Magazine (1 October 2020). www.quantamagazine.org/building-the-mathematical-library-of-the-future-20201001/

45. J. Urban and J. Jakubuv, 'First neural conjecturing datasets and experimenting', *Intelligent Computer Mathematics* (17 July 2020), pp. 315–23.

46. For the original 'proof by exhaustion' of the Kepler conjecture see T. Hales, 'A proof of the Kepler conjecture', *Annals of Mathematics*, vol. 162 [3] (2005), pp. 1065–185. The computer-assisted proof is laid out in T. Hales et al., 'A formal proof of the Kepler conjecture', *Forum of Mathematics, Pi* (29 May 2017).

47. See, for instance, D. Castelvecchi, 'Mathematicians welcome computer-assisted proof in "grand unification" theory', *Nature* (18 June 2021). www.nature.com/articles/d41586–021–01627–2

48. A famous example is the so-called ABC conjecture, which concerns solutions to the equation $a + b = c$. Mathematician Shinichi Mochizuki is one of the few mathematicians actively working on the problem and has become notorious for the sheer complexity of his papers, in which he has invented vast swathes of notation to describe abstract properties of these equations. In an almost comical development, his work was made more 'accessible' when a colleague produced an abridged version, which still came in at 300 pages. See T. Revell, 'Baffling ABC maths proof now has impenetrable 300-page "summary"', *New Scientist* (7 September 2017). www.newscientist.com/article/2146647-baffling-abc-maths-proof-now-has-impenetrable-300-page-summary/

49. M. du Sautoy, *The Creativity Code* (Fourth Estate, 2019), p. 281.

50. William Thurston defines mathematical progress in terms of advancing 'human understanding' of the subject. W. P. Thurston, 'On proof and progress in mathematics', *Bulletin of the American Mathematical Society*, vol. 30 [2] (1994), pp. 161–77.

51. H. Poincaré, 'The future of mathematics', *MacTutor History of Mathematics* (1908/2007). mathshistory.st-andrews.ac.uk/Extras/Poincare_Future/

52. V. Goel et al., 'Dissociation of mechanisms underlying syllogistic reasoning', *Neuroimage*, vol. 12 [5] (2000), pp. 504–14.

53. Erdös's concept was turned into reality posthumously with the publication of M. Aigner and Günter, *Proofs from THE BOOK* (Springer-Verlag, 1998), which shares proofs from a range of topics in mathematics, including number theory, geometry and graph theory. Speaking of their criteria, the authors note that a proof 'can't be too long; it has to be clear; there has to be a special idea; it might connect things that usually one wouldn't think of as having any connection'. See E. Klarreich, 'In search of God's perfect proofs', *Quanta Magazine* (19 March 2018). www.quantamagazine.org/gunter-ziegler-and-martin-aigner-seek-gods-perfect-math-proofs-20180319/

54. A good summary of the criteria for mathematical beauty is found in F. Su, *Mathematics for Human Flourishing* (Princeton University Press, 2020), p. 70. Su includes the listed criteria of Harold Osborne (known for his work on aesthetics): *coherence, lucidity, elegance, clarity, significance, depth, simplicity, comprehensiveness and insight*. Su goes on to distinguish specific types of mathematical beauty. He defines 'insightful beauty' as the beauty of understanding and reasoning, as distinguished from 'sensory beauty', which is concerned with objects. Su goes further still to define 'transcendent beauty' as the deepest kind, where reasoning leads to more profound truths and connections between ideas.

55. S. G. B. Johnson and S. Steinerberger, 'Intuitions about mathematical beauty: a case study in the aesthetic experience of ideas', *Cognition*, vol. 189 (August 2019), pp. 242–59.

56. For example, Christian Szegedy's group at Google Research has examined computer proofs as natural language structures

and attempted to convey them in terms of graphical representations. C. Szegedy et al., 'Graph representations for higher-order logic and theorem proving', *AAAI 2020*, research. google/pubs/pub48827/

The computer scientist Scott Viteri, meanwhile, has identified common features in the structures of selected machine and human-generated proofs: S. Viteri and S. DeDeo, 'Explosive proofs of mathematical truths', *arXiv.org* (31 March 2020). arxiv.org/abs/2004.00055v1

57. O. Roeder, 'An A.I. finally won an elite crossword tournament', *Slate* (27 April 2021). slate.com/technology/2021/04/american-crossword-puzzle-tournament-dr-fill-artificial-intelligence.html

Chapter 4. Imagination

1. See, for example, E. Cheng, *The Art of Logic* (Profile Books, 2018), p. 18.
2. J. Haidt, *The Righteous Mind* (Penguin, 2013).
3. S. Loyd, *Sam Loyd's Cyclopedia of 5000 Puzzles, Tricks and Conundrums with Answers* (Ishi Press, 2007).
4. A brief history of the dots puzzle is presented in R. Eastaway, 'Thinking outside outside the box', *Chalkdust Magazine* (12 March 2018). The success rate for this problem is consistently less than 10 per cent, and often close to zero – see T. C. Kershaw and S. Olsson, 'Multiple causes of difficulty in insight: the case of the nine-dot problem', *Journal of Experimental Psychology: Learning, Memory, and Cognition*, vol. 30 (2004), pp. 3–13. With some aids – such as the strategic placement of two additional dots or a spacious box that fills the nine dots – the success rate increases as subjects are more able to break out of the assumption that the lines must stay within the grid: J. N. MacGregor et al., 'Information processing and insight: a process model of performance on the nine-dot and related problems', *Journal of Experimental Psychology: Learning, Memory, and Cognition*, vol. 27 (2001), pp. 176–201.

For more examples of problems that require adopting a new approach, see the collection of brainteasers studied in V. Goel et al., 'Differential modulation of performance in insight and divergent thinking tasks with tDCS', *Journal of Problem Solving*, vol. 8 (2015).

5. D. R. Hofstadter, *Metamagical Themas* (Basic Books, 1985), p. 47.

6. The term *jootsing* was coined in D. Hofstadter, *Gödel, Escher, Bach: An Eternal Golden Braid* (Basic Books, 1979), and is also one of the 'intuition pumps' in D. Dennett, *Intuition Pumps and Other Tools for Thinking* (Penguin, 2014), pp. 45–48.

7. The mathematics behind several popular video game worlds is detailed in M. Lane, *Power-Up* (Princeton University Press, 2019).

8. T. Kuhn, *The Structure of Scientific Revolutions* (University of Chicago Press, 1962).

9. As quoted in J. Ellenberg, *How Not to Be Wrong* (Penguin, 2014), p. 395.

10. C. Baraniuk, 'For AI to get creative, it must learn the rules – then how to break 'em', *Scientific American* (25 January 2018). www.scientificamerican.com/article/for-ai-to-get-creative-it-must-learn-the-rules-mdash-then-how-to-break-lsquo-em/

11. For a rigorous account of the Pythagoreans' relationship with number, see S. Lawrence and M. McCartney, *Mathematicians and their Gods* (Oxford University Press, 2015), Chapter 2.

12. The origins of zero are far from certain and our best estimates are routinely updated by carbon-dating methods. See T. Revell, 'History of zero pushed back 500 years by ancient Indian text', *New Scientist* (14 September 2017). www.newscientist.com/article/2147450-history-of-zero-pushed-back-500-years-by-ancient-indian-text/

13. This is the English translation, taken from J. Gray, *Plato's Ghost: The Modernist Transformation of Mathematics* (Princeton University Press, 2008), p. 153. Gray attributes the

original German quote to a lecture referenced in H. L. Weber, 'Kronecker', *Jahresbericht der Deutschen Mathematiker-Vereinigun*, 1891–2, p. 19.

14. W. V. Quine, 'The ways of paradox', in W. V. Quine (ed.) *The Ways of Paradox and Other Essays* (Random House, 1966), pp. 3–20. www.pathlms.com/siam/courses/8264/sections/11775/video_presentations/112769

15. T. Aquinas, *Commentary on Aristotle's Physics* (Aeterna Press, 2015).

16. For a particularly imaginative historical account of the formalist endeavour, see the comic A. Doxiadis and C. H. Papadimitriou, *Logicomix* (Bloomsbury, 2009).

17. Taken from V. Kathotia, 'Paradise Lost, Paradise Regained', *Cambridge Mathematics Mathematics Salad* (12 May 2017). www.cambridgemaths.org/blogs/paradise-lost-paradox-regained/

18. As quoted in L. Surette, *The Modern Dilemma* (McGill-Queen's University Press, 2008), p. 340.

19. As quoted in H. W. Eves, *Mathematical Circles Adieu* (Prindle, Weber & Schmidt, 1977).

20. E. Nagel and J. Newman, *Gödel's Proof* (NYU Press, 2001), Chapter VIII.

21. For a summary of Lucas's argument, see J. R. Lucas, 'The implications of Gödel's Theorem', *Etica & Politica*, vol. 5 [1] (2003).

22. R. Penrose, *The Emperor's New Mind* (Oxford University Press, 2016), Chapter 4.

23. For a collection of the most popular objections, see 'The Lucas–Penrose argument about Gödel's Theorem', *Internet Encyclopedia of Philosophy*, iep.utm.edu/lp-argue/

24. P. Benacerraf, 'God, the Devil, and Gödel', *The Monist*, vol. 51 [1] (1967), pp. 9–32.

25. E. Nagel and J. Newman, *Gödel's Proof* (NYU Press, 2001), Foreword.

26. Mathematician Will Byers characterises the reluctance to accept anomalies as a left-hemisphere activity. 'Computation is the preferred mode of the left hemisphere and the left hemisphere is reluctant to accept anomalies or even to admit their existence. Its preference is for consistency and so it views new data through the lens of old schemas or, if that is not possible, it may deny outright the existence of the inconsistent data. Therefore creativity only can occur when the systematic mental picture of the situation breaks down and it is this breakdown that is experienced as painful.' W. Byers, *Deep Thinking* (World Scientific, 2014), p. 123.

27. J. Latson, 'Did Deep Blue beat Kasparov because of a system glitch?', *Time* (17 February 2015). time.com/3705316/deep-blue-kasparov/

28. The historian Johan Huizinga notes that spoilsports often form communities around their newly formed rules, and that their rebellion is rooted in play. See J. Huizinga, *Homo Ludens* (Angelico Press, 2016), p. 12.

Chapter 5. Questioning

1. The origins of this quote are traced by Quote Investigator: quoteinvestigator.com/2011/11/05/computers-useless/

2. A. Turing, 'Computing machinery and intelligence', *Mind*, vol. LIX [236] (1950), pp. 433–60.

3. As estimated in P. Harris, *Trusting What You're Told*, (Harvard University Press, 2015) and reported in L. Neyfakh, 'Are we asking the right questions?', *Boston Globe* (20 May 2012).

4. J. A. Litman et al., 'Epistemic curiosity, feeling-of-knowing, and exploratory behaviour', *Cognition and Emotion*, vol. 19 [4] (2005), pp. 559–282.

5. A survey of definitions of curiosity, and its relation to intrinsic and extrinsic drives, can be found in C. Kidd and B. Y. Hayden, 'The psychology and neuroscience of curiosity', *Neuron*, vol. 88 [3] (2015), pp. 449–60.

6. A discussion of this study, with study references, can be found in S. Baron-Cohen, *The Pattern Seekers* (Allen Lane, 2020), p. 112.

7. G. Loewenstein, 'The psychology of curiosity: a review and reinterpretation', *Psychological Bulletin*, vol. 116 [1] (1994), pp. 75–98.

8. The puzzle is narrated and discussed in the BBC Radio 4 Series *Two Thousand Years of Puzzling*. www.bbc.co.uk/programmes/b09pyrsz

9. This quote can be found among the review pages of several of Martin Gardner's books; for example, M. Gardner, *Did Adam and Eve have Navels?* (W. W. Norton & Company, 2001).

10. A. Bellos, *Puzzle Ninja* (Guardian Faber Publishing, 2017), p. xiv.

11. A similar sentiment is expressed by Tetsuya Miyamoto, a Japanese maths teacher whose KenKen puzzle (similar in style to Nikoli puzzles) has spawned global contests. Miyamoto is adamant that he can tell his own handwritten instances of the puzzle apart from computer-generated ones. See N. Jahromi, 'The Puzzle Inventor Who Makes Math Beautiful', *New Yorker* (30 December 2020). www.newyorker.com/culture/the-new-yorker-documentary/the-puzzle-inventor-who-makes-math-beautiful

12. A. Barcellos, 'A conversation with Martin Gardner', *The Two-Year College Mathematics Journal*, vol. 10 [4] (1979), pp. 233–44.

13. An excellent summary of the story and mathematics behind the Königsberg problem can be found at T. Paoletti, 'Leonhard Euler's solution to the Königsberg bridge problem', *Mathematical Association of America*. The quotes used here are taken from that summary. www.maa.org/press/periodicals/convergence/leonard-eulers-solution-to-the-konigsberg-bridge-problem

14. The letter is the subject of K. Devlin, *The Unfinished Game* (Basic Books, 2010).

15. D. H. Fischer, *Historians' Fallacies* (Harper Perennial, 1970), p. 3.

16. W. T. Gowers, 'The two cultures of mathematics', *www. dpmms.cam.ac.uk/~wtg10/2cultures.pdf*

17. F. Dyson, 'Birds and frogs', *Notices of the American Mathematical Society*, vol. 56 [2] (2009), pp. 212–23.

18. I. Berlin, *The Hedgehog and the Fox* (Princeton University Press, 1953).

19. The metaphor of telescopes and spaceships is taken from James Gleick's foreword to T. Lin, *The Prime Number Conspiracy* (MIT Press, 2018).

20. See computerbasedmath.org and C. Wolfram, *The Math(s) Fix* (Wolfram Media Inc, 2020).

21. T. Vander Ark, 'Stop Calculating and Start Teaching Computational Thinking', *Forbes* (29 June 2020). www.forbes. com/sites/tomvanderark/2020/06/29/stop-calculating-and-start-teaching-computational-thinking/?sh=31c812333786

22. Z. Kleinman, 'Emma Haruka Iwao smashes pi world record with Google help', *BBC News* (14 March 2019). www.bbc. co.uk/news/technology-47524760

23. 'One motivation for computing digits of π is that these calculations are excellent tests of the integrity of computer hardware and software. This is because if even a single error occurs during a computation, almost certainly the final result will be in error. On the other hand, if two independent computations of digits of π agree, then most likely both computers performed billions or even trillions of operations flawlessly.' D. H. Bailey et al., 'The Quest for Pi', *Mathematical Intelligencer*, vol. 19 [1] (1997), pp. 50–57. crd-legacy.lbl. gov/~dhbailey/dhbpapers/pi-quest.pdf

24. 'Swiss researchers calculate pi to new record of 62.8tn figures', *Guardian* (16 August 2021). www.theguardian.com/science/

2021/aug/16/swiss-researchers-calculate-pi-to-new-record-of-628tn-figures

25. D. Castelvecchi, 'AI maths whiz creates tough new problems for humans to solve', *Nature* (3 February 2021). www.nature.com/articles/d41586–021–00304–8

26. For background on the Conway knot, and its unexpected proof, see E. Klarreich, 'Graduate Student Solves Decades-Old Conway Knot Problem', *Quanta Magazine* (19 May 2020). www.quantamagazine.org/graduate-student-solves-decades-old-conway-knot-problem-20200519/

27. As cited in A. Hodges, *Alan Turing: The Enigma* (Vintage, 2012/1983), p. 120.

28. One may be forgiven for thinking that the contrived nature of undecidability renders these examples of undecidable maths problem highly arcane and unworthy of our concern. Yet mathematicians have demonstrated the undecidability of many tangible problems for which they would have expected (and wanted) a proof to exist. See M. Freiberger, 'Picking holes in mathematics', *Plus Magazine* (23 February 2011). plus.maths.org/content/picking-holes-mathematics

29. For an accessible introduction to the *P vs NP* problem see L. Fortnow, *The Golden Ticket* (Princeton University Press, 2017).

30. It has been established, for example, that there are problems that fall outside the NP class but can be solved efficiently by a quantum computer. We can think of such problems as a whole new complexity class of their own. K. Hartnett, 'Finally, a problem that only quantum computers will ever be able to solve', *Quanta Magazine* (21 June 2018). www.quantamagazine.org/finally-a-problem-that-only-quantum-computers-will-ever-be-able-to-solve-20180621/

31. N. Wolchover, 'As math grows more complex, will computers reign?', *Wired* (4 March 2013). www.wired.com/2013/03/computers-and-math/

32. N. Postman, *Building a Bridge to the 18th Century* (Vintage Books, 2011), p. 133.
33. P. Freire, *Pedagogy of the Oppressed* (Penguin, 1993).
34. D. L. Zabelina and M. D. Robinson, 'Child's play: facilitating the originality of creative output by a priming manipulation', *Psychology of Aesthetics, Creativity, and the Arts*, vol. 4 [1] (2010), pp. 57–65.
35. T. Chamorro-Premuzic, 'Curiosity is as important as intelligence', *Harvard Business Review* (27 August 2014). hbr. org/2014/08/curiosity-is-as-important-as-intelligence. The term 'curiosity intelligence' was coined by journalist Thomas Friedman.
36. For an example of how the same dataset can drive different conclusions depending on how the models are set up, see M. Schweinsberg et al., 'Same data, different conclusions: radical dispersion in empirical results when independent analysts operationalize and test the same hypothesis', *Organizational Behavior and Human Decision Processes*, vol. 165 (2021), pp. 228–49.
37. N. Tomašev et al., 'Assessing game balance with AlphaZero: exploring alternate rule sets in chess', *arXiv.org* (15 September 2020).

Chapter 6. Temperament

1. A. Benjamin, 'Faster than a calculator', *TEDx Oxford* (8 April 2013). www.youtube.com/watch?v=e4PTvXtz4GM
2. 'The most powerful computers on the planet', *IBM*. www.ibm. com/thought-leadership/summit-supercomputer/
3. J. Treanor, 'The 2010 "flash crash": How it unfolded', *Guardian* (22 April 2015). www.theguardian.com/business/2015/ apr/22/2010-flash-crash-new-york-stock-exchange-unfolded
4. H. Moravec, *Robot: Mere Machine to Transcendent Mind* (Oxford University Press, 1999), p. 50.
5. Estimates from Mae-Wan Ho, 'The computer aspires to

the human brain', *Science in Society Archive* (13 March 2013) and N. R. B. Martins et al., 'Non-destructive whole-brain monitoring using nanobots: neural electrical data rate requirements', *International Journal of Machine Consciousness*, vol. 4 [1] (2012), pp. 109–40. This paper derives its own estimate of $(5.52 \pm 1.13) \times 10^{16}$ for the brain's electrical data processing rates using an electrophysiological approach.

6. Or, as has been suggested, brain as quantum computer. See G. James, 'Why physicists say your brain might be more powerful than every computer combined', *Inc.* (19 February 2019). www.inc.com/geoffrey-james/why-physicists-say-your-brain-might-be-more-powerful-than-every-computer-combined.html

7. N. Patel, 'Life's too short for slow computers', *Verge* (3 May 2016). www.theverge.com/2016/5/3/11578082/lifes-too-short-for-slow-computers

8. The focus of one of Benjamin's books, *Think Like a Maths Genius,* is well captured by its subtitle: *The Art of Calculating in Your Head*. It is mental maths on steroids; a peek inside exposes vast trenches of abstract algorithms for computing all manner of sums. A. Benjamin and M. Shermer, *Think Like a Maths Genius* (Souvenir Press, 2011).

9. Child Genius: www.channe14.com/programmes/child-genius

10. My accolades are summarised to an almost worrying level of detail on the fan-made Countdown wiki: wiki.apterous.org/Junaid_Mubeen

11. Quoted from www.vedicmaths.org/introduction/what-is-vedic-mathematics, retrieved 16 December 2020.

12. H. S. Bal, 'The fraud of Vedic maths', *Open Magazine* (12 August 2010). www.openthemagazine.com/article/art-culture/the-fraud-of-vedic-maths

13. G. G. Joseph, *The Crest of the Peacock* (Penguin, 1991), pp. 225–39.

14. See, for example: trachtenbergspeedmath.com/

15. In her seminal paper on the topic, Lisanne Bainbridge describes the irony in terms of depleted human performance when we disengage from the tasks that have been outsourced to computers. L. Bainbridge, 'Ironies of automation', *Automatica*, vol. 19 [6] (1983), pp. 775–79.

16. S. Frederick, 'Cognitive Reflection and Decision Making', *Journal of Economic Perspectives*, vol. 19 [4], pp. 25–42.

17. S. Singh, *Pi of Life* (Rowman & Littlefield, 2017), p. 100.

18. As cited in *Infinity and Beyond* (New Scientist: The Collection, 2017), p. 63.

19. R. Webb, 'How to think about … Probability', *New Scientist* (10 December 2014). www.newscientist.com/article/mg22429991–100-how-to-think-about-probability/

20. E. Carey et al., 'Understanding mathematics anxiety: investigating the experiences of UK primary and secondary school students', *Centre for Neuroscience in Education (University of Cambridge)* (March 2019). www.repository.cam.ac.uk/handle/1810/290514

21. S. Beilock, *Choke* (Constable, 2011).

22. T. Lin, *The Prime Number Conspiracy* (MIT Press, 2018), p. 150.

23. T. Gowers, *Mathematics: A Very Short Introduction* (Oxford University Press, 2002), p. 128.

24. H. Poincaré, *The Foundations of Science* (Cambridge University Press, 1913), p. 386.

25. These examples are sourced from M. Popova, 'How Einstein thought: why "combinatory play" is the secret of genius', *Brain Pickings* (14 August 2013). www.brainpickings.org/2013/08/14/how-einstein-thought-combinatorial-creativity/)

26. J. Hadamard, *The Mathematician's Mind* (Princeton University Press, 1996).

27. Quotes from M. Popova, 'French polymath Henri Poincaré on how the inventor's mind works, 1908', *Brain Pickings* (11 June

2012). www.brainpickings.org/2012/06/11/henri-poincare-on-invention/

28. J. Kounios and M. Beeman, 'The cognitive neuroscience of insight', *Annual Review of Psychology*, vol. 65 [1] (2014), pp. 71–93.

29. G. Polya, *How to Solve It* (Princeton University Press, 1971), p. 198.

30. H. E. Gruber, 'On the relation between aha experiences and the construction of ideas', *History of Science Cambridge*, vol. 19 [1] (1981), pp. 41–59. It is also advocated for by Andrew Wiles – see L. Butterfield, 'An evening with Sir Andrew Wiles', *Oxford Science Blog* (30 November 2017). www.ox.ac.uk/news/science-blog/evening-sir-andrew-wiles

31. For a detailed neuroscientific perspective on the cognitive benefits of sleep, see S. Dehaane, *How We Learn* (Penguin, 2019), Chapter 10, and M. Walker, *Why We Sleep* (Penguin, 2018), Chapter 6. For the specific account of how brain wave functions replay events during sleep see E. Renken, 'Dueling brain waves anchor or erase learning during sleep', *Quanta Magazine* (24 October 2019).

32. In one study, researchers tasked a group of Northwestern undergraduates to solve ninety-six Compound Remote Associate problems. You have probably tackled a variant of this exercise, where you are given three words (e.g. crab, pine, sauce) and are asked to find a word that forms a compound word with each of them (e.g. crabapple, pineapple, applesauce). The subjects were also asked to rate the difficulty of each problem and, for the ones they could not solve, to indicate whether they felt the answer was on the tip of their tongue (TOT). The experiment unfolded over two days. To test the effects of sleep, at the end of the first day, after the students had attempted the problems, they were told that they would receive a new set of problems the next day and not to think about problems from the first day any longer. On day two, the researchers gave

the students forty-eight new problems, as well as the original ninety-six problems from the previous day. What they found was that the students, who all professed not to dwell on the previous day's problems as per their instructions, solved a higher percentage of the problems they experienced TOT for. In the words of the authors, 'they [the students] were more likely to solve those problems (compared to problems when they did not experience a TOT) after an overnight incubation period'.

See A. K. Collier and M. Beeman, 'Intuitive tip of the tongue judgments predict subsequent problem solving one day later', *Journal of Problem Solving*, vol. 4 (2012). docs.lib.purdue.edu/jps/v014/iss2/9/

33. T. Lin, *The Prime Number Conspiracy* (MIT Press, 2018), p. 107.

34. As quoted in an interview with Ben Orlin: B. Orlin, 'The state of being stuck', *Math with Bad Drawings blog* (20 September 2017). mathwithbaddrawings.com/2017/09/20/the-state-of-being-stuck/

35. C. Villani, *Birth of a Theorem* (Vintage, 2016).

36. S. Roberts, 'In mathematics, "you cannot be lied to"', *Quanta Magazine* (21 February 2017). www.quantamagazine.org/sylvia-serfaty-on-mathematical-truth-and-frustration-20170221/

37. A summary of Dweck's research into mindset, and its applications, can be found at www.mindsetworks.com/science. For a detailed account see C. Dweck, *Mindset* (Robinson, 2017).

38. A. Duckworth et al., 'Grit: perseverance and passion for long-term goals', *Journal of Personality and Social Psychology*, vol. 92 [6] (2007), pp. 1087–101. See also A. Duckworth, *Grit* (Vermilion, 2017).

39. Plutarch, *The Parallel Lives*, in vol. V of the Loeb Classical Library edition (1917). Retrieved 31 July 2021 from penelope.

uchicago.edu/Thayer/e/roman/texts/plutarch/lives/marcellus*. html

40. M. Knox, 'The game's up: jurors playing Sudoku abort trial', *Sydney Morning Herald* (11 June 2008). www. smh.com.au/news/national/jurors-get-1-million-trial-aborted/2008/06/10/1212863636766.html

41. J. Bennett, 'Addicted to Sudoku', *Newsweek* (22 February 2006). www.newsweek.com/addicted-sudoku-113429

42. M. Csikszentmihalyi, *Flow* (Harper and Row, 1990), p. 4.

43. A. Ericsson and R. Pool, *Peak* (Houghton Mifflin Harcourt, 2016), p. 99.

44. More precisely, Bloom found that students who received one-to-one tutoring performed two standard deviations better than students who learned via conventional instructional techniques (put another way, the average tutored student outperformed 98 per cent of students in the control group). B. Bloom, 'The 2 Sigma problem: the search for methods of group instruction as effective as one-to-one tutoring', *Educational Researcher*, vol. 12 [6] (1984), pp. 4–16.

45. E. Frenkel, *Love and Math* (Basic Books, 2014), p. 56.

46. N. Carr, 'Is Google Making Us Stupid?', *Atlantic* (July 2008). www.theatlantic.com/magazine/archive/2008/07/is-google-making-us-stupid/306868/

47. R. F. Baumeister et al., 'Ego depletion: Is the active self a limited resource?', *Journal of Personality and Social Psychology*, vol. 75 [6] (1998), pp. 1252–65.

48. www.qamacalculator.co.uk

49. N. Bostrom, *Superintelligence* (Oxford University Press, 2014), p. 107.

50. The seminal work on self-determination theory is R. M. Ryan and E. L. Deci, 'Self-determination theory and the facilitation of intrinsic motivation, social development, and well-being', *American Psychologist*, vol. 55 [1] (2000), pp. 68–78.

Chapter 7. Collaboration

1. This biography of Ramanujan, and his collaboration with Hardy, is sourced from R. Kanigel, *The Man Who Knew Infinity* (Abacus, 1992); S. Wolfram, 'Who was Ramanujan?', *Stephen Wolfram Writings (Blog)* (27 April 2016); and E. Klarreich, 'Mathematicians chase moonshine's shadow', in T. Lin (ed.), *The Prime Number Conspiracy* (MIT Press, 2018).

2. Carlyle's 'Great Man' theory is based on two assumptions: a) leadership attributes are predominantly innate and b) great leaders arise when the need for them is greatest. See, for example, B. A. Spector, 'Carlyle, Freud and the Great Man theory more fully considered', *Leadership*, vol. 12 [2] (2016), pp. 250–60.

3. W. E. Wallace, 'Michelangelo, C.E.O.', *New York Times* (16 April 1994). www.nytimes.com/1994/04/16/opinion/michelangelo-ceo.html

4. An informal analysis of 'Great Groups' is undertaken via a series of case studies that includes the Manhattan Project, the Disney studio and Bill Clinton's campaign team – see W. Bennis, *Organizing Genius* (Basic Books, 1998).

5. In A. W. Woolley et al., 'Evidence for a collective intelligence factor in the performance of human groups', *Science*, vol. 330 [6004] (2010), pp. 684–8, the researchers first issued a set of tasks to different groups and found that how a group performed in one task correlated with performance in others, thus indicating the existence of group intelligence (which they denoted c, analogous to g for individual general intelligence). Next, they showed that c is a good predictor of group performance on new tasks, reinforcing the existence of group-level attributes in problem solving.

6. A seminal treatment of emergence in ant colonies is offered in D. Gordon, *Ants at Work* (Free Press, 1999).

7. J. Goldstein, 'Emergence as a construct: history and issues', *Emergence*, vol. 1 [1] (1999), pp. 49–72.

8. The pervasiveness of emergent behaviour has been popularised by S. Johnson, *Emergence* (Penguin, 2002).

9. L. Rozenblit and F. Keil, 'The misunderstood limits of folk science: an illusion of explanatory depth', *Cognitive Science*, vol. 26 [5] (2002), pp. 521–62. As discussed in S. Sloman and P. Fernbach, *The Knowledge Illusion* (Pan, 2018), p. 21.

10. This telling of Galton's ox-weighing experiment is taken from the introduction of J. Surowiecki, *The Wisdom of Crowds* (Abacus, 2005).

11. The insight that learning is encoded in the genome, and reflected in brain structure, is also starting to inform the design of artificial neural networks. See A. M. Zador, 'A critique of pure learning and what artificial neural networks can learn from animal brains', *Nature Communications*, vol. 10 [3770] (2019). www.nature.com/articles/s41467–019–11786–6

12. S. E. Asch, 'Effects of group pressure upon the modification and distortion of judgements', *Swathmore College* (1952), pp. 222–36. www.gwern.net/docs/psychology/1952-asch.pdf

13. I. L. Janis, *Victims of Groupthink: A Psychological Study of Foreign-Policy Decisions and Fiascoes* (Houghton Mifflin, 1972), p. 27. Janis explored the decision-making dynamics that took place during the Cuban missile crisis.

14. J. Surowiecki, *The Wisdom of Crowds* (Abacus, 2004), p. 10. Notably, the book title is a play on *Extraordinary Popular Delusions and the Madness of Crowds*, a classic 1841 work that chronicles the fuzzy end of human collaboration.

15. See C. Pang, *Explaining Humans* (Viking, 2020), p. 52.

16. *Cognitive diversity* is defined as 'differences in perspectives, insights, experiences and thinking styles' in M. Syed, *Rebel Ideas* (John Murray, 2019), Chapter 1.

17. A meta-analysis of 22 studies, covering 5,279 individuals in 1,356 groups, found that Collective Intelligence is a predictor of group performance on problem solving tasks, more so than individual skill and perceptiveness. The same study found that

'The proportion of women in a group is a significant predictor of group performance, mediated by social perceptiveness.' C. Riedl et al., 'Quantifying collective intelligence in human groups', *Proceedings of the National Academy of Sciences*, vol. 118 [21] (2021). www.pnas.org/content/118/21/e2005737118.abstract?etoc

18. A. Saltelli et al., 'Five ways to ensure that models serve society: a manifest', *Nature*, vol. 582 [7813] (2020). www.researchgate.net/publication/342413582_Five_ways_to_ensure_that_models_serve_society_a_manifesto

19. A. Costello, 'The government's secret science group has a shocking lack of expertise', *Guardian* (27 April 2020). www.theguardian.com/commentisfree/2020/apr/27/gaps-sage-scientific-body-scientists-medical

20. For example, www.independentsage.org/

21. Ensemble Covid models are discussed in J. Cepelewicz, 'The hard lessons of modeling the coronavirus pandemic', *Quanta Magazine* (28 January 2021). www.quantamagazine.org/the-hard-lessons-of-modeling-the-coronavirus-pandemic-20210128/

22. For a discussion and some variants of Nisbett's research see L. Winerman, 'The culture–cognition connection', *American Psychological Association*, vol. 37 [2] (2006). www.apa.org/monitor/feb06/connection

23. This study was first carried out by Alexander Luria in Central Asia and is discussed in D. Epstein, *Range* (Macmillan, 2019), pp. 42–4.

24. The 'Flynn effect', named after psychologist James Flynn, is the observation that IQ has increased significantly in more than thirty countries over the past century. The rise is attributed in large part to the increasing exposure that recent generations have had to abstraction and other skills that are aligned to IQ tests. See J. R. Flynn, *Are We Getting Smarter?* (Cambridge University Press, 2012).

25. A figure of 82 per cent is quoted in Kaggle's 'State of machine

learning and data science 2020', which surveyed over 20,000 respondents. Retrieved 1 July 2021 from www.kaggle.com/ kaggle-survey-2020

26. S. M. West et al., 'Discriminating systems: gender, race and power in AI', *AI Now* (April 2019). Retrieved 1 July 2021 from ainowinstitute.org/discriminatingsystems.pdf

27. T. Simonite, 'What really happened when Google ousted Timnit Gebru', *Wired* (8 June 2021). www.wired.com/story/ google-timnit-gebru-ai-what-really-happened/

28. P. Stephan, 'The economics of science', *Journal of Economic Literature*, vol. 34 (1996), pp. 1220–21.

29. E. Wenger et al., *Cultivating Communities of Practice* (Harvard Business Press, 2002), p. 10.

30. J. Love, 'A virtuous mix allows innovation to thrive', *Kellogg Insight* (4 November 2013). insight.kellogg.northwestern.edu/ article/a_virtuous_mix_allows_innovation_to_thrive

31. R. Aboukhalil, 'The rising trend in authorship', *Winnower* (11 December 2014). thewinnower.com/papers/ the-rising-trend-in-authorship

32. T. Hornyak, 'Did Higgs yield the most authors in a single scientific study?', *CNET* (10 September 2012). www.cnet.com/ news/did-higgs-yield-the-most-authors-in-a-science-study/

33. See C. King, 'Multiauthor papers: onward and upward', *ScienceWatch* (July 2012). archive.sciencewatch.com/ newsletter/2012/201207/multiauthor_papers/ and S. Mallapaty, 'Paper authorship goes hyper', *Nature Index* (30 January 2018). www.natureindex.com/news-blog/paper-authorship-goes-hyper

34. J. W. Grossman, 'Patterns of research in mathematics', *Notices of the American Society*, vol. 52 [1] (2005).

35. Posted on MathOverflow: B. Thurston, 'What's a mathematician to do?', *MathOverflow* (30 October 2010). mathoverflow.net/questions/43690/whats-a-mathematician-to-do/44213

36. S. Singh, 'The extraordinary story of Fermat's Last Theorem',

Telegraph (3 May 1997). www.cs.uleth.ca/~kaminski/esferm03. html

37. Cited in R. Elwes, 'An enormous theorem: the classification of finite simple groups', *Plus Magazine* (7 December 2006). plus. maths.org/content/enormous-theorem-classification-finite-simple-groups

38. N. Wiener, *The Human Use of Human Beings* (DaCapo Press, 1988), p. 51.

39. See, for example, S. McChrystal, *Team of Teams* (Penguin, 2015), which transfers principles of military warfare to the corporate context, with emphasis on flat hierarchies that enable faster and more flexible collaboration. A 'team of teams' refers to a collaborative group that contains members of multiple departments.

40. A. McAfee and E. Brynjolfsson, *Machine, Platform, Crowd* (W. W. Norton & Company, 2017), p. 21.

41. M. Haddad, 'Wikipedia Is the last best place on the internet', *Wired* (17 February 2020). www.wired.com/story/wikipedia-online-encyclopedia-best-place-internet/

42. T. Gowers, 'Is massively collaborative mathematics possible?', *Gowers's Weblog* (27 January 2009). gowers.wordpress. com/2009/01/27/is-massively-collaborative-mathematics-possible/

43. M. Nielsen, 'The Polymath project: scope of participation', *Personal Blog* (20 March 2009). michaelnielsen.org/blog/ the-polymath-project-scope-of-participation/

44. See, for example, the CrowdMath project: artofproblemsolving. com/polymath

45. J. Ito, 'Extended intelligence', *MIT Media Lab* (11 February 2016). pubpub.ito.com/pub/extended-intelligence

46. F. Warneken et al., 'Cooperative activities in young children and chimpanzees', *Child Development*, vol. 77 [3] (2006), pp. 640–63.

Epilogue

1. 'The bitter lesson' refers to computer scientist Rich Sutton's view that high levels of computation have proven sufficient to solve some of AI's most difficult problems (that were presumed to require more human-centric approaches like hard-coded knowledge representations). R. Sutton, 'The bitter lesson' (13 March 2019). www.incompleteideas.net/IncIdeas/BitterLesson.html

2. One attempt to estimate the limit of human endurance (measured in terms of the rate of calorie burning) places the cap at two and a half times the body's resting metabolic rate. Higher rates draw on the body's energy stores and are not sustainable longer term. C. Thurber et al., 'Extreme events reveal an alimentary limit on sustained maximal human energy expenditure', *Science Advances*, vol. 5 [6] (2019). advances.sciencemag.org/content/5/6/eaaw0341

3. For example, Y. Harari, *Homo Deus* (Harvill Secker, 2016), Chapter 10.

4. S. Zuboff, *The Age of Surveillance Capitalism* (Profile Books, 2019).

INDEX

Note: The index covers the introduction, the numbered chapters and the epilogue. *Italic* page references indicate a relevant illustration on that page; the suffix 'n', a relevant footnote.